IN SEARCH OF
SCHRÖDINGER'S CAT

Ro. Telford

IN SEARCH OF
SCHRÖDINGER'S CAT

Quantum Physics and Reality

JOHN GRIBBIN

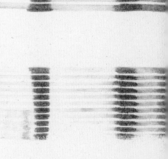

BANTAM BOOKS
Toronto · New York · London · Sydney · Auckland

IN SEARCH OF SCHRÖDINGER'S CAT:
QUANTUM PHYSICS AND REALITY

A Bantam Book / September 1984
6 printings through February 1988

Library of Congress Cataloging-in-Publication Data

Gribbin, John R.
  In search of Schrödinger's cat.

  Bibliography: p. 277
  Includes index.
  I. Quantum theory—History.   2. Reality.
3. Schrödinger, Erwin, 1887–1961.   I. Title.
QC173.98.G75   1984   530.1'2'09   84-2975
ISBN 0-553-34103-0

Published simultaneously in the United States and Canada

Bantam Books are published by Bantam Books, a division of
Bantam Doubleday Dell Publishing Group, Inc. Its trademark,
consisting of the words "Bantam Books" and the portrayal
of a rocster, is Registered in U.S. Patent and Trademark Office
and in other countries. Marca Registrada. Bantam Books,
666 Fifth Avenue, New York, New York 10103.

PRINTED IN THE UNITED STATES OF AMERICA

CW   15   14   13   12   11   10   9   8   7   6

"I don't like it, and I'm sorry
I ever had anything to do with it."

ERWIN SCHRÖDINGER
1887–1961

"Nothing is real."

JOHN LENNON
1940–1980

ACKNOWLEDGMENTS

My acquaintance with quantum theory goes back more than twenty years to my school days, when I discovered the magical way in which the electron-shell model of the atom explained the periodic table of the elements and virtually all of the chemistry that I had struggled with through many a tedious lesson. Following up this discovery for myself with the aid of library books allegedly "too far advanced" for my modest scholastic level, I immediately discovered the beautiful simplicity of the quantum theory's explanation of atomic spectra, and experienced for the first time the revelation that the best things in science are both beautiful and simple, a fact that all too many teachers conceal from their students, by accident or design. I felt like the character in C. P. Snow's *The Search*—which I only read much later—who discovers much the same thing:

> I saw a medley of haphazard facts fall into line and order . . . "But it's true," I said to myself. "It's very beautiful. And it's true." (Macmillan edition, 1963, page 27.)

Partly as a result of this insight, I decided to read physics at university. In due course the ambition was fulfilled, and I became an undergraduate at the University of Sussex in Brighton. But there, the simplicity and beauty of the underlying ideas was smothered in a wealth of detail and mathematical recipes for solving specific problems with the aid of the equations of quantum mechanics. Applying these ideas in the world of physics today seemed to bear as much relevance to the underlying truth and beauty as piloting a 747 must bear to hang gliding, and although the power of that initial insight remained as a major influence in my career, for a long time I neglected the quantum world and explored other scientific pastures.

The fires of that early interest were rekindled by a combination of factors. In the late 1970s and early 1980s, books and articles began to appear attempting, with varying success, to introduce the strange world of the quantum to a nonscientific audience. Some of these alleged "popularizations" were so outrageously far from the truth that I could not imagine any reader discovering the truth and beauty of science by reading them, and I began to feel moved to do the job properly. At the same time, news was coming in of the continuing series of experiments that has now established the reality of some of the strangest features of quantum theory, and that news inspired me to delve back into the libraries and refresh my understanding of those strange ideas. Finally, one Christmas I was asked by the BBC to appear on a radio program as a kind of scientific counterweight to Malcolm Muggeridge, who had recently announced his conversion to the Catholic faith and was the principal guest at the time of this festival. After the great man had had his say, emphasizing the mysteries of Christianity, he turned to me and said "but here's the man who knows all the answers, or claims to know all the answers." In the limited time at my disposal, I endeavored to respond in kind, pointing out that science does *not* claim to have all the answers, and that it is religion, not science, that depends essentially on absolute faith and conviction that the truth is known. "I don't *believe* anything," I said, and was about to expand on this philosophy when the program came to an end. All through the festive season, I was greeted by both friends and acquaintances with an echo of those words, and spent hours explaining that my lack of absolute faith in anything did not prevent me from leading a normal life making use of such reasonable working hypotheses as the likelihood that the sun won't disappear overnight.

The process crystalized my thoughts on what science is all about, and involved a lot of discussion of the basic reality—or unreality—of the quantum world, enough to convince me that I really was ready to write the book you now hold. While preparing the book, I tried out many of the more subtle arguments in my regular scientific contributions to a radio show hosted by Tommy Vance and broadcast by the British Forces Broadcasting Service; Tom's probing questions soon uncovered deficiencies in my presentation, and resulted in a better organization of my ideas. The main source of the reference material used in preparing the book was the library of the University of Sussex, which must have one of the best collections of books on quantum theory anywhere, and some more obscure references were tracked down for me by Mandy Caplin, of *New Scientist,* who has a persuasive way with telex messages, while Christine Sutton straightened out some of my misconceptions about particle physics and field theory. My wife not only provided the essential backup in terms of literary research and organization of the material, but smoothed out many of the rough edges. I am grateful also to Professor Rudolf Peierls for taking the trouble to explain to me in detail some of the subtleties of the "Clock in the Box" experiment and the "EPR Paradox."

So any praise for the good qualities in this book should be laid at the doors of: the "advanced" chemistry texts, whose titles I now forget, that I found in the Kent County Library at the age of sixteen; the misguided "popularizers" and publicists for quantum ideas, who convinced me that I could do it better; Malcom Muggeridge and the BBC; the University of Sussex library; Tommy Vance and BFBS; Mandy Caplin and Christine Sutton; and especially Min. Any complaints concerning the remaining deficiencies in the book should, of course, be addressed to me.

JOHN GRIBBINS
July 1983

CONTENTS

PART ONE

THE QUANTUM

PART TWO

QUANTUM MECHANICS

PART THREE

. . . AND BEYOND

INTRODUCTION

If all the books and articles written for the layman about relativity theory were laid end to end, they'd probably reach from here to the moon. "Everybody knows" that Einstein's theory of relativity is the greatest achievement of twentieth-century science, and everybody is wrong. But if all the books and articles written for the layman about quantum theory were laid end to end, they'd just about cover my desk. That doesn't mean that quantum theory is unheard of outside the halls of academe. Indeed, quantum mechanics has become highly popular in some quarters, being invoked to explain phenomena such as telepathy and spoon bending, and providing a fruitful input of ideas for several science fiction stories. Quantum mechanics is identified in popular mythology, so far as it is identified at all, with the occult and ESP, some weird and esoteric branch of science that nobody understands and nobody has any practical use for.

This book was written to counter that attitude toward what is, in fact, the most fundamental and important area of scientific study. The book owes its genesis to several factors that came together in the summer of 1982. First, I had just finished writing a book about relativity, *Spacewarps*, and felt that it would be appropriate to tackle the demystification of the other great branch of twentieth-century science. Second, I was at that time increasingly irritated by the misconceptions trading under the name quantum theory among some nonscientists, Fritjof Capra's excellent *The Tao of Physics* having spawned imitators who understood neither the physics nor the Tao but suspected there was money to be made out of linking western science with eastern philosophy. Finally, in August 1982 the news came from Paris that a team had successfully carried out a crucial test confirming, for those who still doubted, the accuracy of the quantum-mechanical view of the world.

Don't look here for any "eastern mysticism," spoon bending or ESP. Do look here for the true story of quantum mechanics, a truth far stranger than any fiction. Science is like that—it doesn't need dressing up in the hand-me-downs of someone else's philosophy, because it is full of its own delights, mysteries, and surprises. The question this book addresses is "What is reality?" The answer(s) may surprise you; you may not believe them. But you will find out how contemporary science views the world.

NOTHING IS REAL

The cat of our title is a mythical beast, but Schrödinger was a real person. Erwin Schrödinger was an Austrian scientist instrumental in the development, in the mid-1920s, of the equations of a branch of science now known as quantum mechanics. Branch of science is hardly the correct expression, however, because quantum mechanics provides the fundamental underpinning of all of modern science. The equations describe the behavior of very small objects—generally speaking, the size of atoms or smaller—and they provide the *only* understanding of the world of the very small. Without these equations, physicists would be unable to design working nuclear power stations (or bombs), build lasers, or explain how the sun stays hot. Without quantum mechanics, chemistry would still be in the Dark Ages, and there would be no science of molecular biology—no understanding of DNA, no genetic engineering—at all.

Quantum theory represents the greatest achievement of science, far more significant and of far more direct, practical use than relativity theory. And yet, it makes some very strange predictions. The world of quantum mechanics is so strange, indeed, that even Albert Einstein found it incomprehensible, and refused to accept all of the implications of the theory developed by Schrödinger and his colleagues. Einstein, and many other scientists, found it more comfortable to believe that the equations of quantum mechanics simply represent some sort of mathematical trick, which just happens to give a reasonable working guide to the behavior of atomic and subatomic particles but that conceals some deeper truth that corresponds more closely to our everyday sense of reality. For what quantum mechanics says is that nothing is real and that we cannot say anything about what things are doing when we are not looking at them. Schrödinger's mythical cat was invoked to make the differences between the quantum world and the everyday world clear.

In the world of quantum mechanics, the laws of physics that are familiar from the everyday world no longer work. Instead, events are governed by probabilities. A radioactive atom, for example, might decay, emitting an electron, say; or it might not. It is possible to set up an experiment in such a way that there is a precise fifty-fifty chance that one of the atoms in a lump of radioactive material will decay in a certain time and that a detector will register the decay if it does happen. Schrödinger, as upset as Einstein about the implications of quantum theory, tried to show the absurdity of those implications by imagining such an experiment set up in a closed room, or box, which also contains a live cat and a phial of poison, so arranged that if the radioactive decay does occur then the poison container is broken and the cat dies. In the everyday world, there is a fifty-fifty chance that the cat will be killed, and without looking inside the box we can say, quite happily, that the cat inside is either dead or alive. But now we encounter the strangeness of the quantum world. According to the theory, *neither* of the two possibilities open to the radioactive material, and therefore to the cat, has any reality unless it is observed. The atomic decay has neither happened nor not happened, the cat has neither been killed nor not killed, until we look

inside the box to see what has happened. Theorists who
accept the pure version of quantum mechanics say that the
cat exists in some indeterminate state, neither dead nor
alive, until an observer looks into the box to see how things
are getting on. Nothing is real unless it is observed.

The idea was anathema to Einstein, among others.
"God does not play dice," he said, referring to the theory
that the world is governed by the accumulation of outcomes
of essentially random "choices" of possibilities at the quantum level. As for the unreality of the state of Schrödinger's
cat, he dismissed it, assuming that there must be some underlying "clockwork" that makes for a genuine fundamental reality of things. He spent many years attempting to
devise tests that might reveal this underlying reality at
work but died before it became possible actually to carry
out such a test. Perhaps it is as well that he did not live to
see the outcome of one line of reasoning that he initiated.

In the summer of 1982, at the University of Paris-
South, in France, a team headed by Alain Aspect completed
a series of experiments designed to detect the underlying
reality below the unreal world of the quantum. The underlying reality—the fundamental clockwork—had been given
the name "hidden variables," and the experiment concerned the behavior of two photons or particles of light flying off in opposite directions from a source. It is described
fully in Chapter Ten, but in essence it can be thought of as
a test of reality. The two photons from the same source can
be observed by two detectors, which measure a property
called polarization. According to quantum theory, this property does not exist until it is measured. According to the
hidden-variable idea, each photon has a "real" polarization
from the moment it is created. Because the two photons are
emitted together, their polarizations are correlated with one
another. But the nature of the correlation that is actually
measured is different according to the two views of reality.

The results of this crucial experiment are unambiguous. The kind of correlation predicted by hidden-
variable theory is not found; the kind of correlation predicted by quantum mechanics is found, and what is more,
again as predicted by quantum theory, the measurement

that is made on one photon has an instantaneous effect on the nature of the other photon. Some interaction links the two inextricably, even though they are flying apart at the speed of light, and relativity theory tells us that no signal can travel faster than light. The experiments prove that there is no underlying reality to the world. "Reality," in the everyday sense, is not a good way to think about the behavior of the fundamental particles that make up the universe; yet at the same time those particles seem to be inseparably connected into some indivisible whole, each aware of what happens to the others.

The search for Schrödinger's cat was the search for quantum reality. From this brief outline, it may seem that the search has proved fruitless, since there is no reality in the everyday sense of the word. But this is not quite the end of the story, and the search for Schrödinger's cat may lead us to a new understanding of reality that transcends, and yet includes, the conventional interpretation of quantum mechanics. The trail is a long one, however, and it begins with a scientist who would probably have been even more horrified than Einstein if he could have seen the answers we now have to the questions he puzzled over. Isaac Newton, studying the nature of light three centuries ago, could have had no conception that he was already on the trail leading to Schrödinger's cat.

PART ONE

THE QUANTUM

"Anyone who is not shocked by quantum theory
has not understood it."

NIELS BOHR
1885–1962

CHAPTER ONE

LIGHT

Isaac Newton invented physics, and all of science depends on physics. Newton certainly built upon the work of others, but it was the publication of his three laws of motion and theory of gravity, almost exactly three hundred years ago, that set science off on the road that has led to space flight, lasers, atomic energy, genetic engineering, an understanding of chemistry, and all the rest. For two hundred years, Newtonian physics (what is now called "classical" physics) reigned supreme; in the twentieth century revolutionary new insights took physics far beyond Newton, but without those two centuries of scientific growth those new insights might never have been achieved. This book is not a history of science, and it is concerned with the new physics— quantum physics—rather than with those classical ideas. But even in Newton's work three centuries ago there were already signs of the changes that were to come—not from his studies of planetary motions and orbits, or his famous three laws, but from his investigations of the nature of light.

Newton's ideas about light owed a lot to his ideas about the behavior of solid objects and the orbits of planets. He realized that our everyday experiences of the behavior of objects may be misleading, and that an object, a particle, free from any outside influences must behave very differently from such a particle on the surface of the earth. Here, our everyday experience tells us that things tend to stay in one place unless they are pushed, and that once you stop pushing them they soon stop moving. So why don't objects like planets, or the moon, stop moving in their orbits? Is something pushing them? Not at all. It is the planets that are in a natural state, free from outside interference, and the objects on the surface of the earth that are being interfered with. If I try to slide a pen across my desk, my push is opposed by the friction of the pen rubbing against the desk, and that is what brings it to a halt when I stop pushing. If there were no friction, the pen would keep moving. This is Newton's first law: every object stays at rest, or moves with constant velocity, unless an outside force acts on it. The second law tells us how much effect an outside force—a push—has on an object. Such a force changes the velocity of the object, and a change in velocity is called acceleration; if you divide the force by the mass of the object the force is acting upon, the result is the acceleration produced on that body by that force. Usually, this second law is expressed slightly differently: force equals mass times acceleration. And Newton's third law tells us something about how the object reacts to being pushed around: for every action there is an equal and opposite reaction. If I hit a tennis ball with my racket, the force with which the racket pushes on the tennis ball is exactly matched by an equal force pushing back on the racket; the pen on my desk top, pulled down by gravity, is pushed against with an exactly equal reaction by the desk top itself; the force of the explosive process that pushes the gases out of the combustion chamber of a rocket produces an equal and opposite reaction force on the rocket itself, which pushes it in the opposite direction.

These laws, together with Newton's law of gravity, explained the orbits of the planets around the sun, and the moon around the earth. When proper account was taken of friction, they explained the behavior of objects on the surface of the earth as well, and formed the foundation of mechanics. But they also had puzzling philosophical implications. According to Newton's laws, the behavior of a particle could be exactly predicted on the basis of its interactions with other particles and the forces acting on it. If it were ever possible to know the position and velocity of every particle in the universe, then it would be possible to predict with utter precision the future of every particle, and therefore the future of the universe. Did this mean that the universe ran like clockwork, wound up and set in motion by the Creator, down some utterly predictable path? Newton's classical mechanics provided plenty of support for this deterministic view of the universe, a picture that left little place for human free will or chance. Could it really be that we are all puppets following our own preset tracks through life, with no real choice at all? Most scientists were content to let the philosophers debate that question. But it returned, with full force, at the heart of the new physics of the twentieth century.

WAVES OR PARTICLES?

With his physics of particles such a success, it is hardly surprising that when Newton tried to explain the behavior of light he did so in terms of particles. After all, light rays are observed to travel in straight lines, and the way light bounces off a mirror is very much like the way a ball bounces off a hard wall. Newton built the first reflecting telescope, explained white light as a superposition of all the colors of the rainbow, and did much more with optics, but always his theories rested upon the assumption that light consisted of a stream of tiny particles, called corpuscles. Light rays bend as they cross the barrier between a lighter and a denser substance, such as from air to water or glass (which is why a swizzle stick in a gin and tonic appears to be bent), and this refraction is neatly explained on the corpuscular theory provided the corpuscles move faster in the more "optically dense" substance. Even in Newton's day, however, there was an alternative way of explaining all of this.

The Dutch physicist Christiaan Huygens was a contemporary of Newton, although thirteen years older, having been born in 1629. He developed the idea that light is not a stream of particles but a wave, rather like the waves moving across the surface of a sea or lake, but propagating through an invisible substance called the "luminiferous ether." Like ripples produced by a pebble dropped into a pond, light waves in the ether were imagined to spread out in all directions from a source of light. The wave theory explained reflection and refraction just as well as the corpuscular theory. Although it said that instead of speeding up the

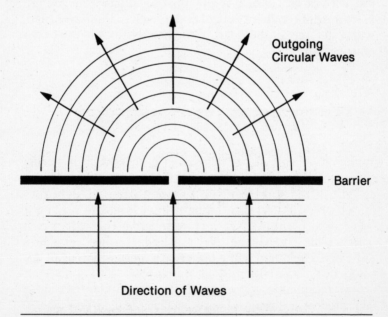

Figure 1.1/Parallel water waves passing through a small hole in a barrier spread out in circles from the hole, leaving no "shadow."

light waves moved more slowly in a more optically dense substance, there was no way of measuring the speed of light in the seventeenth century, so this difference could not resolve the conflict between the two theories. But in one key respect the two ideas did differ observably in their predictions. When light passes a sharp edge, it produces a sharply edged shadow. This is exactly the way streams of particles, traveling in straight lines, ought to behave. A wave tends to bend, or diffract, some of the way into the shadow (think of the ripples on a pond, bending around a rock). Three hundred years ago, this evidence clearly favored the corpuscular theory, and the wave theory, although not forgotten, was discarded. By the early

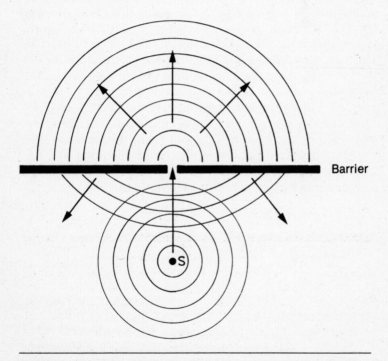

Barrier

Figure 1.2/Circular ripples, like the ones produced by a stone dropped in a pond, also spread as circular waves centered on the hole when they pass through a narrow opening (and, of course, the waves that hit the barrier are reflected back again).

nineteenth century, however, the status of the two theories had been almost completely reversed.

In the eighteenth century, very few people took the wave theory of light seriously. One of the few who not only took it seriously but wrote in support of it was the Swiss Leonard Euler, the leading mathematician of his time, who made major contributions to the development of geometry, calculus and trigonometry. Modern mathematics and physics are described in arithmetical terms, by equations; the techniques on which that arithmetical description depends were largely developed by Euler, and in the process he introduced shorthand methods of notation that survive to this day—the name "pi" for the ratio of the circumference of a circle to its diameter; the letter i to denote the square root of minus one (which we shall meet again, along with pi); and the symbols used by mathematicians to denote the operation called integration. Curiously, though, Euler's entry in the *Encyclopaedia Britannica* makes no mention of his views on the wave theory of light, views which a contemporary said were not held "by a single physicist of prominence."* About the only prominent contemporary of Euler who did share those views was Benjamin Franklin; but physicists found it easy to ignore them until crucial new experiments were performed by the Englishman Thomas Young just at the beginning of the nineteenth century, and by the Frenchman Augustin Fresnel soon after.

WAVE THEORY TRIUMPHANT

Young used his knowledge of how waves move across the surface of a pond to design an experiment that would test whether or not light propagates in the same way. We all know what a water wave looks like, although it is important to think of a ripple, rather than a large breaker, to make the

---

*Quote from page 2 of *Quantum Mechanics,* by Ernest Ikenberry; see bibliography.

analogy accurate. The distinctive feature of a wave is that it raises the water level up slightly, then depresses it, as the wave passes; the height of the crest of the wave above the undisturbed water level is its amplitude, and for a perfect wave this is the same as the amount by which the water level is pushed down as the wave passes. A series of ripples, like the ones from our stone dropped into the pond, follow one another with a regular spacing, called the wavelength, which is measured from one crest to the next. Around the point where our pebble drops into the water, the waves spread out in circles, but the waves on the sea, or ripples produced on a lake by the blowing wind, may run forward as a series of straight lines, parallel waves, one behind the other. Either way, the number of wave crests passing by

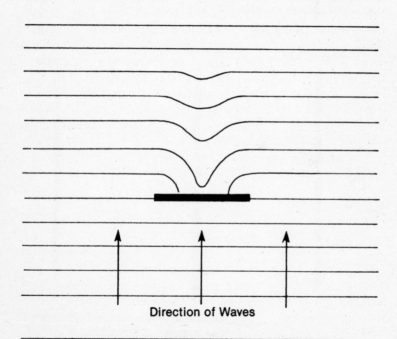

Direction of Waves

*Figure 1.3/*The ability of waves to bend around corners also means that they can quickly fill in the shadow behind an obstacle, provided the obstacle is not much bigger than the wavelength of the waves.

some fixed point—like a rock—in each second tells us the frequency of the wave. The frequency is the number of wavelengths passing each second, so the velocity of the wave, the speed with which each crest advances, is the wavelength multiplied by the frequency.

The crucial experiment starts out with parallel waves, rather like the lines of waves advancing toward a beach before they break. You can imagine these as the waves produced by dropping a very large object into the water a very long way away. The "ripples" spreading out in ever-growing circles look like parallel, or plane, waves if you are far enough away from the source of the ripples, because it is difficult to detect the curvature of the very large circle centered on the spot where the disturbances started. It is easy to investigate what happens to such plane waves in a tank of water when an obstacle is placed in their path. If the obstacle is small, the waves bend around it and fill in be-

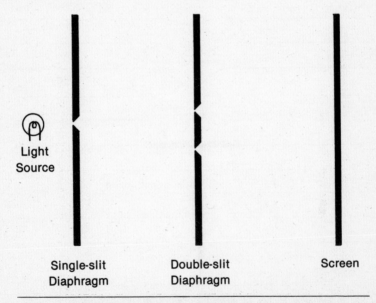

Light
Source

Single-slit   Double-slit   Screen
Diaphragm   Diaphragm

Figure 1.4/The ability of light to diffract around corners and through small holes can be tested using a single slit to make a circular wave and a double slit to produce interference.

hind by diffraction, leaving very little "shadow"; but if the obstruction is very large compared with the wavelength of the ripples, then they only bend slightly into the shadow behind it, leaving a region of undisturbed water. If light is a wave, it is still possible to have sharp-edged shadows, provided the wavelength of light is very small compared with the size of the object casting the shadow.

Now turn the idea around. Imagine a nice set of plane waves progressing across our tank of water and coming up to, not an obstruction surrounded by water but a complete wall across their path, with a gap in the middle. If the gap is much larger than the wavelength of the disturbance, just the portion of the wave that is lined up with the gap gets through, spreading out slightly but leaving most of the water on the other side of the barrier undisturbed—like the waves arriving at the entrance to a harbor wall. But if the hole in the wall is very small, the hole acts as a new source of circular waves, as if pebbles were being dropped into the water at that spot. On the far side of the wall, this

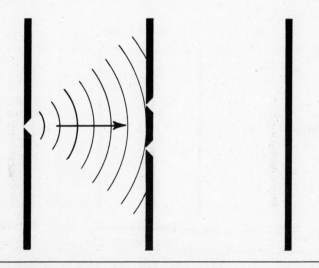

*Figure 1.5/*Like water ripples passing through a hole, the light waves spread out in circles from the first slit, moving "in step" with one another.

circular (or, more accurately, semicircular) wave spreads out across the water surface, leaving no part undisturbed.

So far, so good. Now, at last, we come to Young's experiment. Imagine the same setup as before, a water tank with parallel waves coming to a barrier, but this time a barrier with *two* small holes in it. Each of the holes acts like a new source of semicircular waves in the region of the tank beyond the wall, and because these two sets of waves are being produced by the same parallel waves on the other side of the wall, they move exactly in step, or in phase. Now, we have two sets of ripples spreading out across the water, and this produces a more complicated pattern of ripples on its surface. Where both waves are lifting the water surface upward, we get a more pronounced crest; where one wave is trying to create a crest and the other is trying to create a trough the two cancel out and the water level is undisturbed. The effects are called constructive and destructive interference, and are easy to see, in a cruder way, by dropping two pebbles into a pond at once. If light is a wave, then an equivalent experiment should be able to produce

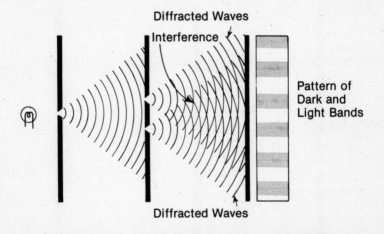

*Figure 1.6/*Circular waves advancing from *each* of the holes in the doubly slitted screen interfere to produce a pattern of light and shade on the viewing screen—clear proof that, as far as this experiment is concerned, light behaves as a wave.

similar interference among light waves, and that is exactly what Young discovered.

He shone a light upon an obstructing screen in which there were two narrow slits. Behind this obstruction, light from the two slits spread out and interfered. If the analogy with water waves was correct, there ought to be a pattern of interference behind the obstruction producing alternate zones of bright light and darkness, caused by constructive and destructive interference of waves from each slit. When Young placed a white screen behind the slits, that is exactly what he found—alternate bands of light and shade striping the screen.

But Young's experiment didn't exactly set the world of science on fire, especially in Britain. The scientific establishment there regarded opposition to any idea of Newton's as almost heretical, and certainly unpatriotic. Newton had only died in 1727, and in 1705—less than a hundred years before Young announced his discoveries—had been the first man to receive a knighthood for his scientific works. It was too soon for the idol to be dethroned in England, so perhaps it was appropriate at that time of the Napoleonic wars that it was a Frenchman, Augustin Fresnel, who took up this "unpatriotic" idea and eventually established the wave explanation of light. Fresnel's work, although a few years later than Young's, was more complete, offering a wave explanation of virtually all aspects of the behavior of light. Among other things, he accounted for a phenomenon familiar to all of us today, the beautifully colored reflections produced when light shines on a thin film of oil. The process is again caused by interference of waves. Some light is reflected from the top of the film of oil, but some passes through and is reflected back from the bottom surface of the layer. So there are two different reflected beams, which interfere with one another. Because each color of light corresponds to a different wavelength, and white light is composed of a superposition of all the colors of the rainbow, the reflections of a white light from an oil film will produce a mass of colors as some waves (colors) interfere destructively and some constructively, depending on just where your eye is in relation to the film.

By the time Léon Foucault, the French physicist famous for the pendulum that bears his name, established in the middle of the nineteenth century that, contrary to the predictions of Newton's corpuscular theory, the speed of light is less in water than in air, this was no more than any reputable scientist expected. By then "everybody knew" that light was a form of wave motion propagated through the ether, whatever that might be. Still, however, it would be nice to know exactly what is "waving" in a beam of light. In the 1860s and 1870s, the theory of light seemed at last to have been completed when the great Scottish physicist James Clerk Maxwell established the existence of waves involving changing electric and magnetic fields. This electromagnetic radiation was predicted by Maxwell to involve patterns of stronger and weaker electric and magnetic fields in the same way that water waves involve crests and troughs in the height of the water. In 1887—only a hundred years ago—Heinrich Hertz succeeded in transmitting and receiving electromagnetic radiation in the form of radio waves, which are similar to light waves but have much longer wavelengths. At last the wave theory of light was complete—just in time to be overturned by the greatest revolution in scientific thinking since the time of Newton and Galileo. By the end of the nineteenth century, only a genius or a fool would have suggested that light is corpuscular. His name was Albert Einstein; but before we can understand why he took this bold step we need a little more background concerning the ideas of nineteenth-century physics.

CHAPTER TWO

ATOMS

Many popular accounts of the history of science say that the idea of atoms goes back to the ancient Greeks, a time of the birth of science, and go on to praise the ancients for their early perception of the true nature of matter. But this account is a bit of an exaggeration. It is true that Democritus of Abdera, who died sometime close to 370 B.C., did propose that the complex nature of the world could be explained if all things were composed of different kinds of unchangeable atoms, each type with its own shape and size, in constant motion. "The only existing things are atoms and empty space; all else is mere opinion," he wrote,* and later Epicurius of Samos and the Roman Lucretius Carus adopted the idea. But it was not in those days the front-runner among theories to account for the nature of the world, and Aristotle's suggestion that everything in the universe is made up from the four "elements" fire,

---

*Quoted in many books, including *Invitation to Physics* by Jay M. Pasachoff & Marc L. Kutner (page 3).

earth, air, and water proved much more popular and enduring. While the idea of atoms was largely forgotten by the time of Christ, Aristotle's four elements were accepted for two thousand years.

Although the Englishman Robert Boyle used the concept of atoms in his work on chemistry in the seventeenth century, and Newton had it in mind in his work on physics and optics, atoms only really became a part of scientific thought in the latter part of the eighteenth century when the French chemist Antoine Lavoisier investigated why things burn. Lavoisier identified many real elements, pure chemical substances that cannot be separated into other chemical substances, and he realized that burning is simply the process by which oxygen from the air combines with other elements. In the early years of the nineteenth century John Dalton put the role of atoms in chemistry on a secure footing. He stated that matter is made up of atoms, which are themselves indivisible; that all the atoms of one element are identical, but that different elements have different kinds of atoms (different sizes or shapes); that atoms cannot be created or destroyed, but are rearranged by chemical reactions; and that a chemical compound, made from two or more elements, is composed of molecules, each of which has a small, fixed number of atoms from each of the elements in the compound. So the atomic concept of the material world really came into being, in the form that is taught in textbooks today, less than two hundred years ago.

NINETEENTH-CENTURY ATOMS

Even so, the idea only slowly gained favor among chemists in the nineteenth century. Joseph Gay-Lussac established by experiment that when two gaseous substances combine together the volume of one gas required is always a simple proportion to the volume of the other. If the compound produced is also a gas, the volume of that third gas is also in a simple proportion to the other two. This fits in with the idea that each molecule of the compound is made of one or two atoms of one gas combined with a few atoms of the other. And the Italian Amadeo Avogadro used this evidence, in 1811, to derive his famous hypothesis, which says that for any fixed temperature and pressure equal volumes of gas contain the same number of molecules, whatever the chemical nature of the gas. Later experiments have established that Avogadro's hypothesis is correct; it can be proved that each liter of gas at a pressure of one atmosphere and temperature of 0°C contains roughly 27,000 billion billion (27×10^{21}) molecules. But it was only in the 1850s that Avogadro's countryman, Stanislao Cannizzaro, developed the idea to a point where more than a few chemists began to take it seriously. As late as the 1890s, however, there were still many chemists who did not accept the ideas of Dalton and Avogadro. But by then they had been overtaken by events in the development of physics, where the behavior of gases had been explained in detail, using the concept of atoms, by the Scot James Clerk Maxwell and the Austrian Ludwig Boltzmann.

During the 1860s and 1870s, these pioneers developed the idea that a gas is made up of very many atoms or molecules (the number derived from Avogadro's hypothesis gives you some idea how many), which can be thought of as tiny, hard spheres that bounce around, colliding with one another and with the walls of the container that holds the gas. This related directly to the idea that heat is a form of motion—when a gas is heated, the molecules move faster, which increases the pressure on the walls of the container, and if the walls are not fixed in place, the gas will expand. The key feature of these new ideas was that the behavior of a gas could be explained by applying the laws of mechanics—Newton's laws—in a statistical sense to a very large number of atoms or molecules. Any one molecule might be moving in any direction in the gas at any time, but the combined effect of many molecules colliding with the walls of the container each second produces a steady pressure. This led to the development of a mathematical description of gas processes called statistical mechanics. But still there was no direct proof that atoms existed; some leading physicists of the time argued strongly against the atomic hypothesis, and even in the 1890s Boltzmann felt himself (perhaps mistakenly) to be an individual struggling against the tide of scientific opinion. In 1898, he published his detailed calculations in the hope "that, when the theory of gases is again revived, not too much will have to be rediscovered";[*] in 1906, ill and depressed, unhappy about the continuing opposition of many leading scientists to this kinetic theory of gases, he killed himself, unaware that a few months before an obscure theorist called Albert Einstein had published a paper that established the reality of atoms beyond reasonable doubt.

EINSTEIN'S ATOMS

This paper was just one of three published by Einstein in the same volume of the *Annalen der Physik* in 1905, any one of which would have assured him of a place in the annals of science. One of the papers introduced the special theory of relativity and is largely outside the scope of the present book; another concerned the interaction of light with electrons and was later recognized as the first scientific work dealing with what we now call quantum mechanics—it was for this work that Einstein received the Nobel Prize in 1921. The third paper was a deceptively simple explanation of a puzzle that had baffled scientists since 1827—an explanation that established, as far as any theoretical paper ever could, the reality of atoms.

Einstein later said that his major aim at that time was "to find facts which would guarantee as much as possible the existence of atoms of finite size,"[†] an aim that perhaps indicates the importance of the work at the beginning of

---

[*]Quoted in *The Historical Development of Quantum Theory*, volume one, page 16, by Jagdish Mehra and Helmut Rechenberg.
[†]Quote from Einstein's "Autobiographical Notes," in *Albert Einstein: Philosopher Scientist*, edited by P. A. Schilpp, Tudor, New York, 1949 (page 47).

the present century. At the time these papers were published, Einstein was working as a patent examiner in Berne—his unconventional approach to physics had not made him an obvious candidate for an academic post when he completed his formal education, and the patent office job suited him. His logical mind proved well able to sort out the wheat of new inventions from the chaff, and his skill at the job left him plenty of free time in which to think about physics, even during office hours. Some of his thoughts concerned the discoveries made by the British botanist Thomas Brown almost eighty years before. Brown noticed that when a pollen grain floating in a drop of water is examined using a microscope it is seen to bounce around in an irregular fashion, moving in a random pattern that is now called Brownian motion. Einstein showed that this motion, although random, obeys a definite statistical law, and that the pattern of behavior is exactly what should be expected if the pollen grain is being repeatedly "kicked" by unseen, submicroscopic particles that move in accordance with the statistics used by Boltzmann and Maxwell to describe the way atoms move in a gas or liquid. It looks so obvious today that it is hard to credit what a breakthrough this paper made. You or I, used to the idea of atoms, can see at once that if pollen grains are being jostled by unseen collisions then it must be moving atoms that push them around. But before Einstein made the point, respected scientists could still find room to doubt the reality of atoms; after his paper appeared, there was no longer room to doubt. Easy when explained, like the fall of an apple from a tree, but if it was so obvious why had it not been appreciated in the previous eight decades?

It's ironic that this scientific paper should have been published in German (in the journal *Annalen der Physik*), because it was the opposition of leading German-speaking scientists such as Ernst Mach and Wilhelm Ostwald that seems to have convinced Boltzmann that his was a lone voice crying in the wilderness. In fact, by the beginning of the twentieth century there was a great deal of evidence for the reality of atoms, even if, strictly speaking, that evidence could only be described as circumstantial; British and

French physicists subscribed to the atomic theory with far more conviction than many of their German counterparts, and it was an Englishman, J. J. Thomson, who had discovered the electron—which we now know to be one of the components of the atom—in 1897.

ELECTRONS

There had been a long controversy in the late nineteenth century concerning the nature of the radiation that is produced from a wire that carries an electric current through a tube that has been pumped empty of air. These cathode rays, as they were called, might be a form of radiation, produced by vibrations of the ether but different in character from light and the newly discovered radio waves, or they might be streams of tiny particles. Most German scientists subscribed to the idea of ether waves; most British and French scientists thought that the cathode rays must be particles. The situation was confused by the accidental discovery of X rays by Wilhelm Röntgen in 1895 (in 1901, Röntgen received the first ever Nobel Prize in Physics for this discovery), but that proved to be a red herring. Important though the discovery was, in a sense it came too soon, before there was a theoretical framework of atomic physics into which X rays could be fitted. We will meet them again in a more logical context as our story develops.

Thomson worked at the Cavendish Laboratory, a research center in Cambridge established by Maxwell, as the first Cavendish Professor of Physics, in the 1870s. He devised an experiment that depended on balancing the electric and magnetic properties of a moving charged particle.*

---

*"Devised" is exactly the right word here. J. J. Thomson was notoriously clumsy and planned brilliant experiments that other people carried out; his son George is reported as saying that although J. J. (as he was always known) "could diagnose the faults of an apparatus with uncanny accuracy it was just as well not to let him handle it." (See *The Questioners,* Barbara Lovett Cline, page 13.)

The path of such a particle can be deflected both by magnetic and by electric fields, and Thomson's apparatus was designed so that these two effects canceled each other out and left a beam of cathode rays that traveled straight from a negatively charged metal plate (or cathode) to a detector screen. This trick only works for electrically charged particles; so Thomson established that cathode rays are indeed negatively charged particles (now called electrons*) and he was able to use the balance of electrical and magnetic forces to calculate the ratio of the electric charge of an electron to its mass (e/m). Whatever metal was used to make the cathode, he always got the same result, and concluded that electrons are parts of atoms, but that although different elements are made of different atoms all atoms contain identical electrons.

This was no serendipitous discovery, like that of X rays, but the result of careful planning and skilful experiments. Maxwell had created the Cavendish Laboratory, but it was under Thomson that it became a leading center of experimental physics—perhaps *the* leading physics lab in the world—at the heart of the discoveries that led to the new understanding of physics in the twentieth century. As well as his own prize, Nobel awards went to seven of the people who worked under Thomson at the Cavendish in the period before 1914. It remains a world center of physics to the present day.

IONS

The cathode rays, produced by the negatively charged plate in an evacuated tube, turned out to be negatively charged particles, electrons. Atoms, however, are electrically neu-

---

*The screen on which you see your TV picture is part of just such a tube, called a cathode-ray tube; the cathode rays that paint your TV picture are electrons, swept across the screen by changing magnetic fields just like the ones studied by Thomson.

tral, and logically enough there are positively charged counterparts to electrons, atoms that have had a piece of negative charge chipped away from them. Wilhelm Wien, of the University of Würzburg, made some of the first studies of these positive rays in 1898, and established that the particles they are made of are very much heavier than electrons, as we would expect if they are just atoms that lack an electron. Following his work on cathode rays, Thomson took up the challenge of investigating these positive rays in a series of difficult experiments that extended into the 1920s. Today the rays are called ionized atoms, or simply "ions"; in Thomson's day they were called canal rays, and he studied them by using a modified cathode-ray tube in which a little gas was left behind by the vacuum pump. The electrons moving through this gas collided with its atoms, and knocked other electrons out of them, leaving the positively charged ions, which could be manipulated with electric and magnetic fields in the same way that Thomson manipulated the electrons themselves. By 1913 Thomson's team was making measurements of the deflections of positive ions of hydrogen, oxygen, and other gases. One of the gases Thomson used in these experiments was neon; a trace of neon in an evacuated tube through which an electric current passes glows brightly, and Thomson's apparatus was a forerunner of the modern neon tube. What he found, however, was far more important than a new kind of advertising sign.

Unlike the electrons, which all have the same e/m, it turns out that there are three different neon ions, which all have the same amount of charge as the electron (but $+e$ instead of $-e$), and have different masses. This was the first evidence that chemical elements often include atoms with different masses (different atomic weights) but identical chemical properties. Such variations on an elemental theme are now called "isotopes," but it was a long time before an explanation of their existence could be found. Already, though, Thomson had enough information to produce the first stab at an explanation of what the atom might be like inside—not an indivisible ultimate particle, as a few Greek philosophers had thought, but a mixture of positive and negative charges, from which electrons could be chipped away.

Thomson envisaged the atom as something like a watermelon, a relatively large sphere through which all of the positive charge was spread, with small electrons embedded in it like seeds, each carrying its own little bit of negative charge. He turned out to be wrong, but he literally gave scientists a target to shoot at, and their target practice led to a more accurate understanding of atomic structure. To see how, we have to take one step backward into scientific history, then two steps forward.

X RAYS

The key to unlock the secret of the structure of the atom proved to have been the discovery of radioactivity in 1896. Like the discovery of X rays a few months earlier, this was largely a lucky accident, although in both cases the kind of lucky accident bound to happen in some physics lab around that time. Like many physicists in the 1890s, Wilhelm Röntgen was experimenting with cathode rays. It happens that when these rays—electrons—strike a material object the collision produces a secondary radiation. This radiation is invisible and can only be detected by its effect on photographic plates or film, or on a piece of apparatus called a fluorescent screen, which produces sparks of light when struck by the radiation. Röntgen just happened to have a fluorescent screen lying on a table near his cathode-ray experiment, and he was quick to notice that when the discharge tube in the cathode-ray experiment was operating this screen fluoresced. This led him to discover the secondary radiation, which he called "X" because x is traditionally the unknown quantity in a mathematical equation. The X rays were soon shown to behave like waves (we now know that they are a form of electromagnetic radiation, very similar to light waves but with much shorter wavelengths), and this discovery in a German lab helped to confirm the view of

most German scientists that the cathode rays must also be
waves.

The discovery of X rays was announced in December
1895, and caused a stir in the scientific community. Other
researchers tried to find other ways of producing X rays or
related forms of radiation, and the first to succeed was
Henri Becquerel, working in Paris. The most intriguing
feature of X-radiation was the way that it could pass unim-
peded through many opaque substances, such as black paper, to produce an image on a photographic plate that had
not been exposed to light. Becquerel was interested in
phosphorescence, which is the emission of light by a substance that has previously absorbed light. A fluorescent
screen, like the one featured in the discovery of X rays,
emits light only when it is being "excited" by incoming radiation; a phosphorescent substance has the ability to store
up incoming radiation and release it as light, slowly fading
away, for hours after it has been placed in the dark. It was
natural to seek a relationship between phosphorescence
and X-radiation, but what Becquerel discovered was as unexpected as the discovery of X rays had been.

RADIOACTIVITY

In February 1896 he wrapped a photographic plate in a
double thickness of black paper, coated the paper with
bisulphate of uranium and potassium, and exposed the
whole thing to the sun for several hours. When the plate
was developed, it showed the outline of the coating of
chemicals. Becquerel thought that X-radiation had been
produced in the coating—a uranium salt—by the sunlight,
in the same way as phosphorescence. Two days later, he
prepared another plate in the same way to repeat the experiment, but the sky was cloudy on that day and the next,
and the prepared plate stayed shut away in a cabinet. On 1
March, Becquerel developed that plate anyway, and again
found the outline of the uranium salt. Whatever it was that

had fogged the two plates had nothing to do with sunlight or phosphorescence, but was a previously unknown form of radiation coming, it turned out, from the uranium itself, spontaneously, without any outside influence. This ability to emit radiation spontaneously is now called radioactivity.

Alerted by Becquerel's discovery, other scientists took up the investigation of radioactivity, and Marie and Pierre Curie, working at the Sorbonne, soon became the experts in this new branch of science. For their work on radioactivity and the discovery of new radioactive elements they were awarded the Nobel Prize in Physics in 1903; in 1911 Marie received a second Nobel Prize, in chemistry, for her further work with radioactive material (Irene, the daughter of Marie and Pierre Curie, also received a Nobel Prize for her work on radioactivity in the 1930s). In the early 1900s, experimental discoveries in radioactivity ran far ahead of theory, with a cascade of new developments that were only later incorporated into the theoretical framework. Throughout this period, one name stood out in the investigation of radioactivity, that of Ernest Rutherford.

Rutherford was a New Zealander who had worked with Thomson at the Cavendish in the 1890s. In 1898 he was appointed Professor of Physics at McGill University in Montreal, where he and Frederick Soddy showed in 1902 that radioactivity involves the transformation of the radioactive element into another element. It was Rutherford who found that two different types of radiation are produced by this radioactive "decay," as it is now called, and he gave them the names alpha and beta radiation. When a third type of radiation was discovered later, it was only natural to call it gamma radiation. Both alpha and beta radiation turned out to be fast-moving particles; the beta rays were soon shown to be electrons, the radioactive equivalent of cathode rays, and in due course gamma rays were shown to be another form of electromagnetic radiation, like X rays but with even shorter wavelengths. The alpha particles, however, turned out to be something altogether different—particles with mass about four times that of a hydrogen atom and an electric charge twice as large as the charge on the electron but positive instead of negative.

INSIDE THE ATOM

Even before anyone knew exactly what an alpha particle was, or how it could be ejected at very high speed from an atom that in the process was transformed into an atom of another element, researchers such as Rutherford were able to make use of them. Such high-energy particles, themselves the product of atomic reactions, could be used as probes to study the structure of atoms and, in a curiously circular piece of scientific investigation, find out just where the alpha particles had come from in the first place. In 1907 Rutherford moved from Montreal to become Professor of Physics at the University of Manchester, England; in 1908 he received the Nobel Prize in Chemistry for his work on radioactivity, an award that caused him some wry amusement. Although the study of the elements was regarded by the Nobel Committee as chemistry, Rutherford regarded himself as a physicist and had little time for chemistry, which he regarded as a very inferior branch of science. (With the new understanding of atoms and molecules that is provided by quantum physics, of course, the old joke of physicists that chemistry is simply a branch of physics has become much more than half true.) In 1909 Hans Geiger and Ernest Marsden, working in Rutherford's department at Manchester, carried out experiments in which a beam of alpha particles was directed onto and through a thin foil of metal. The alpha particles came from naturally radioactive atoms—there were no artificial particle accelerators available in those days. The fate of the particles directed onto the metal foil was determined with scintillation counters, fluorescent screens that sparked when struck by such a particle. Some of the particles went straight through the metal foil; some were deflected and came out at an angle to the original beam; and some, to the surprise of the experimenters, bounced back from the foil on the same side that the beam hit it. How could this happen?

Rutherford came up with the answer. Each alpha particle has a mass more than 7,000 times that of an electron (in fact, an alpha particle is identical to a helium atom from which two electrons have been removed) and may be moving at close to the speed of light. If such a particle collides with an electron, it brushes the electron aside and continues unaffected. The deflections must be caused by the positive charges in the atoms of the metal foil (like charges, as with like magnetic poles, repel one another), but if Thomson's watermelon model were correct no particles would be bounced back. If the sphere of positive charge filled the atom, then the alpha particles must go straight through it, since the experiment showed that most of the particles went straight through the foil. If the watermelon let one particle through, it ought to let them all through. But if all of the positive charge were concentrated in a tiny volume, much smaller than that of the whole atom, then just occasionally an alpha particle hitting this tiny concentration of matter and charge head-on would be bounced back, while most of the alpha particles whizzed through the empty space between the positively charged parts of the atoms. Only with this kind of arrangement could the positive charge of the atom sometimes repel the positively charged alpha particles back in their tracks, sometimes deflect them slightly from their path, and sometimes leave them almost undisturbed.

So, in 1911, Rutherford proposed a new model of the atom, one that turned out to be the basis of our modern understanding of atomic structure. He said that there must be a small central region of the atom, which he called the nucleus. The nucleus contains all of the positive charge of the atom, an amount exactly equal and opposite to the amount of negative charge in the cloud of electrons that surround the nucleus, so that nucleus and electrons together make up an electrically neutral atom. Later experiments showed that the size of the nucleus is only about one hundred thousandth of the size of an atom—a nucleus typically about 10^{-13} cm across embedded in an electron cloud typically 10^{-8} cm across. To put these figures in perspective, imagine a pinhead, perhaps a millimeter across, at the center of St. Paul's cathedral, surrounded by a cloud of microscopic dust motes far out in the dome of the cathedral, say 100 meters away. The pinhead represents the atomic nucleus; the dust motes are its retinue of electrons. That is how much empty space there is in the atom—and all of the seemingly solid objects in the material world are made of these empty spaces, held together by electric charges. Rutherford had already won a Nobel Prize, remember, when he came up with this new model of the atom (a model based on experiments he had devised). But his career was hardly over, for in 1919 he announced the first artificial transmutation of an element, and in the same year succeeded J. J. Thomson as Director of the Cavendish Laboratory. He was first knighted (in 1914) and then, in 1931, made Baron Rutherford of Nelson. In spite of all this, including the Nobel Prize, his greatest contribution to science was undoubtedly the nuclear model of the atom. This model was to transform physics, leading as it did to an obvious question—since unlike charges attract each other as eagerly as like charges repel, why don't the negative electrons fall into the positive nucleus? The answer came from an analysis of the way atoms interact with light, and it marked the coming of age of the first version of quantum theory.

CHAPTER THREE

LIGHT AND ATOMS

The puzzle posed by Rutherford's model of the atom depended on the known fact that a moving electric charge that is accelerated radiates energy in the form of electromagnetic radiation—light, radio waves, or some other variation on the theme. If an electron just sat outside the nucleus of an atom, it ought to fall into the nucleus, so the atom would not be stable. As the atom collapsed, it would radiate a burst of energy. The obvious way to counteract this tendency for an atom to collapse was to imagine the electrons orbiting around the nucleus, like the planets orbiting around the sun in our solar system. But orbital motion involves continuous acceleration. The speed of the orbiting particle may not change, but the direction it is moving in does, and both speed and direction together define velocity, which is what matters. As the velocity of orbiting electrons changed, they ought to radiate energy, and as they lost energy as a result they ought to spiral into the nucleus. Even by invoking orbital motion, theorists could not prevent the collapse of Rutherford's atom.

When this model was improved, the theorists started out from the picture of electrons orbiting the nucleus, and tried to find ways to hold them in place in their orbits without losing energy and spiraling inward. This was a natural starting point, fitting in nicely with the obvious analogy of the solar system. But it was wrong. As we shall see, it makes just as much sense to think of the electrons as sitting out in space at some distance from the nucleus, not orbiting around it. The problem is the same—how to stop the electrons falling in—but the image this conjures up is very different from the image of planets orbiting around the sun, and that is entirely a good thing. The trick used by theorists to explain why the electrons don't fall is the same whether or not we use the orbital analogy, and that analogy is both redundant and misleading. The picture most people still have, from school or popular accounts, is of an atom rather like the solar system, with a tiny central nucleus around which electrons whiz in circular orbits. This is the place to abandon that picture, however, and to try to approach the bizarre world of the atom—the world of quantum mechanics—with an open mind. Think simply of nucleus and electrons sitting together in space, and ask why the attraction between positive and negative charges does not cause the atom to collapse, radiating energy as it does so.

By the time theorists began to tackle this puzzle, in the second decade of the twentieth century, the crucial discoveries that were to give them an improved model of the atom had already been made. They depended on studies of the way matter (atoms) interacts with radiation (light).

At the beginning of the twentieth century, the best scientific view of the natural world required a dualistic philosophy. Material objects might be described in terms of particles, or atoms, but electromagnetic radiation, including light, had to be thought of in terms of waves. So the study of the way light and matter interact seemed to provide the best chance of unifying physics in the years around 1900. But it was exactly in the attempt to explain how radiation interacts with matter that classical physics, so successful in almost every other way, broke down.

The simplest way to see (literally) how matter and radiation interact is to look at a hot object. A hot object radiates electromagnetic energy, and the hotter it is the more energy it radiates, at shorter wavelengths (higher frequencies). So a red-hot poker is cooler than a white-hot poker, and a poker that is too cool to radiate visible light may still feel warm, because it radiates lower-frequency infrared radiation. Even at the end of the nineteenth century it was fairly obvious that this electromagnetic radiation must be associated with the movement of tiny electric charges. The electron itself had only just been discovered, but it is easy to see how a charged part of an atom (which we would now identify with an electron) vibrating to and fro will produce a stream of electromagnetic waves, in a manner not too unlike the way you can make water ripples by wiggling a finger to and fro in your bath. The trouble was that a combination of the best classical theories—statistical mechanics, and electromagnetism—predicted a form of radiation very different from the kind actually observed coming from hot objects.

THE BLACKBODY CLUE

In order to make such predictions, theorists used, as they always do, an imaginary idealized example, in this case a "perfect" absorber or emitter of radiation. Such an object is usually called a "blackbody," because it absorbs all the radiation that falls upon it. This is an unfortunate choice of name, however, since it turns out that a blackbody is also the most efficient at turning heat energy into electromagnetic radiation—a "blackbody" could very well be red- or white-hot, and in some ways the surface of the sun itself acts rather like a blackbody. Unlike many of the theorists' idealized conceptions, however, it is easy to make a blackbody in the laboratory. Just take a hollow sphere, or a tube with closed ends, and make a small hole in the side. Any radiation, such as light, which happens to go in through the hole will be trapped inside, bounced around the walls of the container until it is absorbed; it is very unlikely that it will ever just happen to bounce out through the hole, so the hole is, in effect, a blackbody. This gives the radiation its alternative, German name: cavity radiation.

We are more interested, though, in what happens to a blackbody when it is heated. Just like your poker, it first feels warm and then glows red- or white-hot, depending on its temperature. The spectrum of the radiation emitted— the amount radiated at each wavelength—can be studied in the laboratory, by looking at what comes out of a small hole in the side of a hot container, and such studies show that it depends only on the temperature of the blackbody. There is very little radiation at very short wavelengths (high frequencies), and very little at very long wavelengths, with most energy radiated in some middle band of frequencies. The peak of the spectrum shifts toward shorter wavelengths as the body gets hotter (from infrared, to red, to blue, to ultraviolet), but there is always a cutoff at very short wavelengths. This is where the measurements of blackbody radiation made during the nineteenth century came into conflict with theory.

Strange though it sounds, the best predictions of classical theory agreed that a cavity full of radiation should always have an infinite amount of energy at the shortest wavelengths—that instead of a peak in the blackbody spectrum and a falling away to zero energy at zero wavelength, the measurements should go off the scale at the short wavelength end. The calculations came from the seemingly natural assumption that the electromagnetic waves of the radiation in the cavity could be treated in the same way as the waves on a string, such as a violin string, and that there can be waves of any size—any wavelength or frequency. Because there are very many wavelengths (many "modes of vibration") to consider, the laws of statistical mechanics have to be taken over from the world of particles to the world of waves in order to predict the overall appearance of the radiation in the cavity, and this leads directly to the conclusion that the energy radiated at any frequency is proportional to the frequency. Frequency is just the inverse of wavelength, and very short wavelengths are very high frequencies. So all blackbody radiation should produce huge amounts of high-frequency energy, in the ultraviolet and beyond. The higher the frequency, the greater the energy. This prediction is called the "ultraviolet catastrophe," and it shows that there must be something wrong with the assumptions from which the prediction is built.

But all is not quite lost. On the low-frequency side of the blackbody curve, the observations agree very well with the predictions based on classical theory, known as the Rayleigh-Jeans Law. At least classical theory is half right. The puzzle is why the energy of the oscillations at high frequencies is not very large, but actually falls away to zero as the frequency of the radiation gets bigger.

This puzzle attracted the attention of many physicists in the last decade of the nineteenth century. One of them was Max Planck, a German scientist of the old school. Thorough and hard working, Planck was at heart a scientific conservative, not a revolutionary. His special interest was in thermodynamics, and his great hope at that time was to resolve the ultraviolet catastrophe by the application of thermodynamic rules. In the late 1890s two approximate equations were known that between them gave a rough representation of the blackbody spectrum. An early version of the Rayleigh-Jeans Law worked at long wavelengths, and Wilhelm Wien had developed a formula that fitted the observations approximately at short wavelengths, and also "predicted" the wavelength at which the peak of the curve occurred at any temperature. Planck started out by looking at how small electric oscillators ought to radiate and absorb electromagnetic waves, a different approach from the one used by Rayleigh in 1900 and Jeans a little later, but one which gave exactly the standard curve complete with its ultraviolet catastrophe. From 1895 to 1900, Planck worked on the problem and published several key papers that established the connection between thermodynamics and electrodynamics—but still he could not solve the riddle of the blackbody spectrum. In 1900 he made the breakthrough, not through a cool, calm and logical scientific insight, but as an act of desperation mixing luck and insight with a fortunate misunderstanding of one of the mathematical tools he was using.

Of course, no one today can be absolutely sure what was in Planck's mind when he took the revolutionary step that led to quantum mechanics, but his work has been studied in detail by Martin Klein of Yale University, a historian specializing in the history of physics at the time of the birth of quantum theory. Klein's reconstruction of the parts played by Planck and Einstein in that birth is as authentic an account as we are ever likely to get, and it puts the discoveries in a convincing historical context. The first step, in the late summer of 1900, owed nothing to luck and everything to the insight of a trained mathematical physicist. Planck realized that the two incomplete descriptions of the blackbody spectrum could be combined in one simple mathematical formula that described the shape of the whole curve—in effect, he used a little mathematical juggling to bridge the gap between the two formulae, Wien's Law and the Rayleigh-Jeans Law. This was a great success. Planck's equation agreed beautifully with observations of cavity radiation. But unlike the two half-laws from which it was built, it had no physical basis. Wien and Rayleigh— even Planck in the previous four years—had tried to build a theory starting from sensible physical assumptions and leading up to the blackbody curve. Now, Planck had pulled the right curve out of the hat, but nobody knew what physical assumptions "belonged" with that curve. It turned out that they were not very "sensible" at all.

AN UNWELCOME REVOLUTION

Planck's formula was announced at a meeting of the Berlin Physical Society in October 1900. For the next two months he immersed himself in the problem of finding a physical basis for the law, trying out different combinations of physical assumptions to see which ones matched the mathematical equations. He later said that this was the most intensive period of work of his entire life. Many attempts failed, until at last Planck was left with only one, to him unwelcome, alternative.

I described Planck as a physicist of the old school, and so he was. In his earlier work he had been reluctant to accept the molecular hypothesis, and he particularly abhorred the idea of a statistical interpretation of the property known as entropy, an interpretation introduced by Boltzmann into the science of thermodynamics. Entropy is a key concept in physics, related in a fundamental sense to the flow of time. Although the simple laws of mechanics—Newton's Laws—are completely reversible as far as time is concerned, we know that the real world just isn't like that. Think of a stone dropped on the ground. When it hits the ground, the energy of its motion is converted into heat. But if we put an identical stone on the ground and warm it by the same amount, it doesn't jump up into the air. Why not? In the case of the falling stone, an orderly form of motion (all the atoms and molecules falling in the same direction) is turned into a disordered form of motion (all the atoms and molecules jostling against one another energetically but randomly). This is in accordance with a law of nature that seems to require that disorder is always increasing, and disorder is identified, in this sense, with entropy. The law is the second law of thermodynamics, and states that natural processes always move toward an increase of disorder, or that entropy always increases. If you put disordered heat energy into a stone it cannot, in that case, use that energy to create an orderly movement of all the molecules in the stone so that they jump upward together.

Or can it? Boltzmann introduced a variation on the theme. He said that such a remarkable occurrence *could* happen, but it is extremely unlikely. In the same way, as a result of the random movement of air molecules it *could* happen that all of the air in the room might suddenly concentrate in the corners (it has to be more than one corner because the molecules are moving in three space dimensions); but, again, such a possibility is so unlikely that for all practical purposes it can be ignored. Planck argued against this statistical interpretation of the second law of thermodynamics long and hard, both publicly and in correspondence with Boltzmann. For him, the second law was an absolute; entropy *must* always increase, and probabilities didn't enter into it. So it is easy to understand how Planck must have felt near the end of 1900, when, having exhausted all other options he reluctantly tried to incorporate Boltzmann's statistical version of thermodynamics into his calculations of the blackbody spectrum, and found that they worked. The irony of the situation is made more piquant, however, by the fact that because of his unfamiliarity with Boltzmann's equations, Planck applied them inconsistently. He got the right answer, but for the wrong reason, and it wasn't until Einstein took up the idea that the real significance of Planck's work became clear.

It is worth stressing that it was already a major step forward in science for Planck to establish that Boltzmann's statistical interpretation of entropy increase is the best description of reality. Following Planck's work, it could never really be doubted that entropy increase, while very probable indeed, cannot be taken as an absolute certainty. This has interesting implications in cosmology, the study of the whole universe, where we deal with vast stretches of time and space. The bigger the region we deal with, the more scope there is for unlikely things to happen someplace, and sometime inside it. It is even possible (though still not very likely) that the whole universe, which is an orderly place, by and large, represents some sort of thermodynamic statistical fluctuation, a very large, very rare hiccup that has created a region of low entropy that is now running down. Planck's "mistake," however, revealed something even more fundamental about the nature of the universe.

Boltzmann's statistical approach to thermodynamics involved cutting energy up into chunks, mathematically, and treating the chunks as real quantities that could be handled by the probability equations. The energy divided up into portions before this part of the calculation then has to be added together (integrated) at a later stage, to give the total energy—in this case, the energy corresponding to blackbody radiation. Halfway through this procedure, however,

Planck realized that he already had the mathematical formula he was looking for. Before getting to the stage of integrating the pieces of energy back into a continuous whole, the blackbody equation was there in the mathematics. So he took it. This was a very drastic step, and totally unjustified within the context of classical physics.

Any good classical physicist starting out from Boltzmann's equations to construct a blackbody radiation formula would have completed the integration. Then, as Einstein was later to show, the adding up of the pieces of energy would have restored the ultraviolet catastrophe—indeed, Einstein pointed out that *any* classical approach to the problem inevitably brings about this catastrophe. It was only because Planck knew the answer that he was looking for that he was able to stop short of the full, seemingly correct, classical solution of the equations. As a result, he was left with pieces of energy that had to be explained. He interpreted this apparent division of electromagnetic energy into individual pieces as meaning that the electric oscillators inside the atom could only emit or absorb energy in lumps of a certain size, called quanta. Instead of dividing the available amount of energy up in an infinite number of ways, it could only be divided into a finite number of pieces among the resonators, and the energy of such a piece of radiation (E) must be related to its frequency (denoted by the Greek letter nu, v) according to a new formula,

$$E = hv$$

where h is a new constant, now called Planck's constant.

WHAT IS h?

It is easy to see how this resolves the ultraviolet catastrophe. For very high frequencies, the energy needed to emit one quantum of radiation is very large, and only a few of the oscillators will have this much energy (in accordance with the statistical equations) so only a few high-energy quanta are emitted. At very low frequencies (long wavelengths),

very many low-energy quanta are emitted, but they each
have so little energy that even added together they don't
amount to much. Only in the middle range of frequencies
are there plenty of oscillators that have enough energy to
emit radiation in moderate-sized lumps, which add together to produce the peak in the blackbody curve.

But Planck's discovery, announced in December 1900,
raised more questions than it answered, and it failed to set
the world of physics on fire. Planck's own early papers on
the quantum theory are not models of clarity (which perhaps reflects the confused way he was forced to introduce
the idea into his beloved thermodynamics), and for a long
time many—even most—physicists who knew about his
work still regarded it simply as a mathematical trick, a device to get rid of the ultraviolet catastrophe that had little or
no physical significance. Planck himself was certainly confused. In a letter to Robert William Wood written in 1931 he
looked back at his work of 1900 and said, "I can charac-
terize the whole procedure as an act of despair . . . a theoretical interpretation *had* to be found at any price, however
high it might be."* Yet he knew he had stumbled upon
something significant, and according to Heisenberg,
Planck's son later told how his father described his work at
the time, during a long walk through the Grunewald in the
suburbs of Berlin, explaining that the discovery might rank
with those of Newton.†

Physicists were busy in the early 1900s absorbing the
new discoveries involving atomic radiation, and Planck's
new "mathematical trick" to explain the blackbody curve
didn't seem of overwhelming importance alongside those
discoveries. Indeed, it took until 1918 for Planck to receive
the Nobel Prize for his work, a very long time compared
with the speed with which the work of the Curies or
Rutherford was recognized. (This was partly because it always takes longer to recognize dramatic new theoretical
breakthroughs; a new theory isn't as tangible as a new particle, or an X ray, and it has to stand the test of time and be

---

*Quoted by Mehra & Rechenberg, volume 1.
†See *Physics and Philosophy*, page 35.

confirmed by experiments before it achieves full recognition.) There was also something odd about Planck's new
constant, h. It is a very small constant, 6.6×10^{-34} joule
seconds, but that is not so puzzling since if it were much
bigger then it would have made its presence obvious long
before physicists began to puzzle over blackbody radiation.
No, the strange thing about h is the units in which it is
measured, energy (ergs) multiplied by time (seconds).
Such units are called "actions," and they were not an everyday feature of classical mechanics—there is no "law of conservation of action" to rank with the law of conservation of
mass or energy. But an action has one particularly interesting property, which it shares, among other things, with entropy. A constant action is absolutely constant and has the
same size for all observers in space and time. It is a four-
dimensional constant, and the significance of that only became apparent when Einstein unveiled his theory of relativity.

Because Einstein is the next actor to enter upon the
quantum mechanical stage, it might be worth a small diver-
sion to see what this means. The special theory of relativity
treats the three dimensions of space and one of time as a
four-dimensional whole, the space-time continuum. Observers moving through space at different speeds get a different view of things—they will disagree, for example, on
the length of a stick that they measure as it passes. But the
stick can be thought of as existing in four dimensions, and
as it moves "through" time it traces out a four-dimensional
surface, a hyper-rectangle whose height is the length of the
stick and whose breadth is the amount of time that has
passed. The "area" of that rectangle is measured in units of
length × time, and that area comes out to be the same for
all observers who measure it, even though they disagree
with one another about the length and the time they are
measuring. In the same way action (energy × time) is a
four-dimensional equivalent of energy, and action is seen to
be the same for all observers, even when they disagree
about the size of the energy and time components of the
action. In special relativity, there *is* a law of conservation of
action, and it is every bit as important as the law of con-

servation of energy. Planck's constant only looked peculiar because it was discovered before the theory of relativity.

And that emphasizes, perhaps, the holistic nature of physics. Of Einstein's three great contributions to science published in 1905 one, special relativity, seems to be very different from the others, on Brownian motion and the photoelectric effect. Yet they all hang together on the framework of theoretical physics, and in spite of the publicity generated by his theory of relativity the greatest of Einstein's contributions was his work on quantum theory, which jumped off from Planck's work by way of the photoelectric effect.

The revolutionary aspect of Planck's work in 1900 was that it showed a limitation to classical physics. It doesn't matter exactly what that limitation is. Just the fact that there are phenomena that cannot be explained solely with the classical ideas built from the work of Newton was enough to herald a new era in physics. The original form of Planck's work was, however, much more limited than it often seems from modern accounts. There is a school of adventure writing in which the hero escapes miraculously from cliff-hanger situations at the end of every episode, summed up by the phrase "with one bound, Jack was free." Many popular accounts of the birth of quantum mechanics read like scientific one-bound-Jackery. "At the end of the nineteenth century, classical physics had run into a brick wall. With one bound Planck invented the quantum, and physics was free." Far from it. Planck only suggested that the electric oscillators inside atoms might be quantized. He meant that they could only emit packets of energy of certain sizes, because something inside them prevented them from absorbing or emitting "in between" amounts of radiation.

The automatic teller at my London bank works in much the same way. When I put in my cash card, the machine will deliver any amount of money I want, provided that it is a multiple of £5. The automatic teller can't deliver in-between amounts (and it can't deliver less than £5), but that doesn't mean that the in-between amounts, such as £8.47, don't exist. So Planck himself did not suggest that

the *radiation* was quantized, and he seems always to have
been wary of the deeper implications of quantum theory. In
later years, as quantum theory progressed, Planck made
some contributions to the science he had founded, but
spent most of his working life trying to reconcile the new
ideas with classical physics. It wasn't that he had changed
his mind, but rather that he never appreciated in the first
place just how far his blackbody equation was removed
from classical physics—he derived the equation by combining thermodynamics with electrodynamics, and both of
those were classical theories. Rather than having second
thoughts, Planck's efforts to find a halfway house between
quantum ideas and classical physics actually represented a
profound shift, for him, away from the classical ideas on
which he had been brought up. But his grounding in classical ideas was so thorough that it is no surprise to find the
real progress being made by a new generation of physicists,
less set in their ways and less committed to old ideas, fired
by the new discoveries in atomic radiation and looking for
new answers to both old and new questions.

EINSTEIN, LIGHT, AND QUANTA

Einstein was twenty-one in March 1900. He took up his
famous job with the Swiss patent office in the summer of
1902, and in those early years of the twentieth century devoted most of his scientific attention to problems of thermodynamics and statistical mechanics. His first scientific
publications were as traditional in style and in the problems
they tackled as those of the previous generation, including
Planck. But in the first paper he published that refers to
Planck's ideas about the blackbody spectrum (it was published in 1904), Einstein began to break new ground, and
to develop a style of solving physical puzzles that was all his
own. Martin Klein describes how Einstein was the first per-

son to take the physical implications of Planck's work seriously, and to treat them as more than a mathematical trick;* within a year, this acceptance of the equations as having a foundation in physical reality had led to a dramatic new insight, the revival of the corpuscular theory of light.

The other jumping-off point for his 1904 paper, as well as Planck's work, was the investigation of the photoelectric effect by Phillip Lenard and J. J. Thomson, working independently, at the end of the nineteenth century. Lenard, born in 1862 in the part of Hungary that is now Czechoslovakia, received the Nobel Prize for physics in 1905 for his research on cathode rays. Among those experiments, he had shown in 1899 that cathode rays (electrons) can be produced by light shining onto a metal surface in a vacuum. Somehow, the energy in the light makes electrons jump out of the metal.

Lenard's experiments involved beams of light of a single color (monochromatic light) which means that all the waves in the light have the same frequency. He looked at the way the intensity of the light affected the way electrons were ripped out of the metal, and found a surprising result. Using a brighter light (he actually moved the same light closer to the metal surface, which has the same effect) there is more energy shining on each square centimeter of the metal surface. If an electron gets more energy, then it ought to be knocked out of the metal more rapidly, and fly off with a greater velocity. But Lenard found that as long as the wavelength of the light stayed the same all of the ejected electrons flew off with the same velocity. Moving the light closer to the metal increased the number of electrons that were ejected, but each of those electrons still

---

*See Klein's contribution to *Some Strangeness in the Proportion*, edited by Harry Woolf. In the same volume, Thomas Kuhn, of MIT, goes even further than most authorities in arguing that Planck "had no conception of a discrete energy spectrum when he presented the first derivations of his blackbody distribution law" and that Einstein was the first to appreciate "the essential role of quantization in blackbody theory." Kuhn says that "it is Einstein rather than Planck who first quantized the Planck oscillator." That debate we can leave to the academics; but there is no doubt that Einstein's contributions were crucial to the development of quantum theory.

came out with the same velocity as the ones produced by a weaker beam of light of the same color. On the other hand, the electrons *did* move faster when he used a beam of light with a higher frequency—ultraviolet, say, instead of blue or red light.

There is a very simple way to explain this, provided you are prepared to abandon the ingrained ideas of classical physics and take Planck's equations as physically meaningful. The importance of those provisos is clear from the fact that in the five years after Lenard's initial work on the photoelectric effect and Planck's introduction of the concept of quanta nobody took that seemingly simple step. In effect, all Einstein did was to apply the equation $E = h\nu$ to the electromagnetic radiation, instead of to the little oscillators inside the atom. He said that light is *not* a continuous wave, as scientists had thought for a hundred years, but instead comes in definite packets, or quanta. All the light of a particular frequency ν, which means of a particular color, comes in packets that have the same energy E. Every time one of these light quanta hits an electron, it gives it the same amount of energy and therefore the same velocity. More intense light simply means that there are more light quanta (we now call them photons) that all have the same energy, but changing the color of the light changes its frequency, and so changes the amount of energy that is carried by each photon.

This was the work for which Einstein eventually received the Nobel Prize, in 1921. Once again, a theoretical breakthrough had to wait for full recognition. The idea of photons did not gain immediate acceptance, and although the experiments of Lenard agreed with the theory in a general way, it took more than a decade for the exact prediction of the relationship between the velocity of the electrons and the wavelength of the light to be tested and proved. That was achieved by the American experimenter Robert Millikan, who along the way established a very precise measurement of the value of h, Planck's constant. In 1923 Millikan in turn received the Nobel Prize in Physics, for this work and his accurate measurements of the size of the charge on the electron.

So Einstein had a pretty busy year. One paper that led
to the award of the Nobel Prize; another that proved once
and for all the reality of atoms; a third that saw the birth of
the theory for which he is best known: relativity. And, almost incidentally, at the same time, in 1905, he was in the
throes of completing another little piece of work concerning
the size of molecules, which he submitted as his doctoral
thesis to the University of Zurich. The doctorate itself was
awarded in January 1906. Although the PhD was not then
the key to a life of active research that it is today, it is still
remarkable that the three great papers of 1905 were published by a man who could at that time only sign himself
"Mr." Albert Einstein.

In the next few years Einstein continued to work on the
integration of Planck's quantum into other areas of physics.
He found that the idea explained long-standing puzzles
concerning the theory of specific heat (the specific heat of a
substance is the amount of heat needed to raise the temperature of a fixed amount of material by a chosen degree;
it depends on the way atoms vibrate inside the material,
and those vibrations turn out to be quantized). This is a less
glamorous area of science, often overlooked in accounts of
Einstein's work, but the quantum theory of matter gained
acceptance more quickly than Einstein's quantum theory
of radiation, and began the persuasion of many physicists of
the old school that quantum ideas had to be taken seriously.
Einstein refined his ideas on quantum radiation over the
years up until 1911, establishing that the quantum structure of light is an inevitable implication of Planck's equation, and pointing out to an unreceptive scientific world
that the way to a better understanding of light would involve a fusion of the wave and particle theories that had
vied with each other since the seventeenth century. By
1911, his thoughts were turning to other things. He had
convinced himself that quanta were real, and his own opinion was all that mattered. His new interest was the problem
of gravity, and over the five years up to 1916 he developed
his General Theory of Relativity, the greatest of all his
works. It took until 1923 for the reality of the quantum
nature of light to be established beyond all doubt, and this

in turn led to a new debate about particles and waves that helped to transform quantum theory and ushered in the modern version of the theory, quantum mechanics. More of those ideas in their place. The first flowering of quantum theory came in the decade during which Einstein turned away from the subject and concentrated on other matters. It came from a fusion of his ideas with Rutherford's model of the atom, and it came largely as a result of the work of a Danish scientist, Niels Bohr, who had been working with Rutherford in Manchester. After Bohr produced his model of the atom, nobody could doubt any longer the value of quantum theory as a description of the physical world of the very small.

CHAPTER FOUR

BOHR'S ATOM

By 1912 the pieces of the atomic puzzle were ready to be fitted together. Einstein had established the broad validity of the idea of quanta, and had introduced the idea of photons even though this was not yet generally accepted. Extending the cash-dispenser analogy, what Einstein said was that energy really does only come in packets of a definite size—the automatic teller only deals in units of £5 because that is the smallest denomination of currency that there is, not because of some whim on the part of the programmer who set up the machine. Rutherford had produced a new picture of the atom, with a small central nucleus and a surrounding cloud of electrons, though again this idea had yet to gain general support. Rutherford's atom, however, simply could not be stable according to the classical laws of electrodynamics. The solution was to use *quantum* rules to describe the behavior of electrons within atoms. And once again the breakthrough came from a young researcher with a fresh approach to the problem—a continuing theme throughout the story of the development of quantum theory.

Niels Bohr was a Danish physicist who completed his doctorate in the summer of 1911 and went to Cambridge that September to work with J. J. Thomson at the Cavendish. He was a very junior researcher, shy and speaking imperfect English; he found it difficult to find a niche in Cambridge, but on a visit to Manchester he met Rutherford and found him very approachable and interested in Bohr and his work. So in March 1912 Bohr moved up to Manchester and began to work with Rutherford's team, concentrating on the puzzle of the structure of the atom.* After six months he returned to Copenhagen, but only briefly, and he remained associated with Rutherford's group in Manchester until 1916.

JUMPING ELECTRONS

Bohr had a particular genius, which was just the thing required to make progress in atomic physics over the next ten to fifteen years. He didn't worry about explaining all the details in a complete theory, but was quite willing to patch together different ideas to make an imaginary "model" that worked in at least rough agreement with the observation of real atoms. Once he had a rough idea of what was going on, he could tinker with it to make the bits fit together even better and in this way work toward a more complete picture. So he took the image of an atom as a miniature solar system, with electrons moving around orbits in accordance with the laws of classical mechanics and electromagnetism, and he said that the electrons could not spiral inward out of those orbits, emitting radiation as they did so, because they were only allowed to emit whole pieces of en-

---

*One version of the story has it that the move was a result of a disagreement between Bohr and Thomson about Thomson's model of the atom, which Bohr didn't like, and that J. J. quietly suggested that Rutherford might be more receptive to Bohr's ideas. See E. U. Condon, quoted by Max Jammer on page 69 of *The Conceptual Development of Quantum Mechanics*.

ergy—whole quanta—not the continuous radiation required by classical theory. The "stable" orbits of the
electrons corresponded to certain fixed amounts of energy,
each a multiple of the basic quantum, but there were no in-
between orbits because they would require fractional
amounts of energy. Pushing the solar system analogy rather
more than is justified, this is like saying that the Earth's
orbit around the sun is stable, and so is that of Mars, but
that there is no such thing as a stable orbit anywhere in
between.

What Bohr did had no right to work. The whole idea of
an orbit depends on classical physics; the idea of electron
states corresponding to fixed amounts of energy—energy
levels, as they came to be called—comes from quantum
theory. Making a model of the atom by patching together
bits of classical theory and bits of quantum theory gave no
true insight into what made atoms tick, but it did indeed
provide Bohr with enough of a working model to make
progress. His model turns out to have been wrong in almost
every respect, but it provided a transition to a genuine
quantum theory of the atom, and as such it was invaluable.
Unfortunately, because of its nice, simple blend of quantum and classical ideas, and the seductive picture of the
atom as a miniature solar system, the model has outstayed
its welcome in the pages not just of popularizations but of
many school and even university texts. If you learned anything about atoms at school, I'm sure you learned about
Bohr's model, whether or not it was given that name in
class. I won't tell you to forget everything you were told, but
prepare yourself to be persuaded that it was not the whole
truth. And you *should* try to forget the idea of electrons as
little "planets" circling around the nucleus—it's the idea
Bohr had at first, but it really is misleading. An electron is
simply something that sits outside the nucleus and has a
certain amount of energy and other properties. It moves, as
we shall see, in a mysterious way.

The great early triumph of Bohr's work, in 1913, was
that it successfully explained the spectrum of the light from
hydrogen, the simplest atom. The science of spectroscopy

goes back to the early years of the nineteenth century, when William Wollaston discovered dark lines in the spectrum of light from the sun, but it was only with Bohr's work that it came into its own as a tool for probing the structure of the atom. Like Bohr mixing classical and quantum theories to make progress, however, we have to take a step back from Einstein's ideas about light quanta to appreciate how spectroscopy works. In this kind of work, it makes no sense to think of light as anything except an electromagnetic wave.*

White light, as Newton established, is made up of all the colors of the rainbow, the spectrum. Each color corresponds to a different wavelength of light, and by using a glass prism to spread white light out into its colored components we are in effect spreading out the spectrum so that the waves of different frequencies lie beside one another on a screen, or on a photographic plate. Short-wavelength blue and violet light is at one end of the optical spectrum, and long-wavelength red at the other—at both ends, though, the spectrum extends far beyond the range of colors visible to our eyes. When the sun's light is spread out in this way, the spectrum revealed is marked by very sharp, dark lines at very precise places in the spectrum, corresponding to very precise frequencies. Without knowing how these lines were formed, researchers such as Joseph Fraunhofer, Robert Bunsen (whose name is immortalized in the standard laboratory burner) and Gustav Kirchhoff, working in the nineteenth century, established by experiments that each element produces its own set of spectral lines. When an element (such as sodium) is heated in the flame of a bunsen burner, it produces light with a characteristic color (in this case, yellow), which is produced by strong emission of radiation as a bright line or lines in one part of the spectrum. When white light passes through a liquid or gas containing the same element, even if the element is combined with others in a chemical compound, the spectrum in the light shows dark absorption lines, like those in light from

---

*The full quantum theory tells us that light is both particle *and* wave, but we haven't got there yet.

the sun, at the same frequencies characteristic of that element.

This explained the dark lines in the solar spectrum. They must be produced by cooler clouds of material in the atmosphere of the sun, absorbing radiation at the characteristic frequencies from light passing through them from the much hotter solar surface. And the technique gave chemists a useful means of identifying the elements present in a compound. Throw common salt on a fire, for example, and the fire flares with the characteristic sodium-yellow color (a color also familiar today from sodium-yellow street lamps). In the laboratory, the characteristic spectrum can be seen by dipping a wire into the substance being tested and holding it in the flame of a bunsen burner. Every element gives its own pattern of lines, and in each case the pattern stays the same, although its intensity changes, even if the temperature of the flame changes. The sharpness of each spectral line shows that every atom of the element is emitting or absorbing precisely at the same frequency, with none out of step. By comparison with such flame tests, spectroscopists accounted for most of the lines in the spectrum of sunlight, and explained them as due to the presence of elements known on earth. In one famous reversal of this procedure, the English astronomer Norman Lockyer (who founded the scientific journal *Nature*) discovered lines in the solar spectrum that could not be explained in terms of the spectrum of any known element, and said that they must be due to a previously unknown element, which he called helium. In due course, helium was found on earth, and proved to have exactly the spectrum required to fit the solar lines.

With the aid of spectroscopy, astronomers can probe the distant stars and galaxies to find out what they are made of. And atomic physicists can now probe the inner structure of the atom using the same tool.

The spectrum of hydrogen is particularly simple, which we now know is because hydrogen is the simplest element and each atom contains just one positively charged proton for its nucleus, and one negatively charged electron associated with it. The lines in the spectrum that provide the unique fingerprint of hydrogen are called the Balmer lines, after Johann Balmer, a Swiss schoolteacher who worked out a formula describing the pattern in 1885, which just happens to have been the year Niels Bohr was born. Balmer's formula relates the frequencies in the spectrum at which hydrogen lines occur to one another. Starting with the frequency of the first hydrogen line, in the red part of the spectrum, Balmer's formula gives the frequency of the next hydrogen line, in the green. Starting from the green line, the same formula applied to that frequency gives the frequency of the next line, in the violet, and so on.* Balmer only knew about four hydrogen lines in the visible spectrum when he worked out his formula, but other lines had already been discovered and fitted it exactly; when more hydrogen lines were identified in the ultraviolet and infrared, they too fitted this simple numercial relationship. Obviously, the Balmer formula said something significant about the structure of the hydrogen atom. But what?

Balmer's formula was common knowledge among physicists, part of every physics undergraduate course, by the time Bohr came on the scene. But it was just part of a mass of complicated data on spectra, and Bohr was no spectroscopist. When he began working on the puzzle of the structure of the hydrogen atom, he didn't immediately think of the Balmer series of lines as an obvious key to use to unlock the mystery, but when a colleague who specialized in spectroscopy pointed out to him just how simple the Balmer formula really was (regardless of the complexities of the spectra of other atoms) he was quick to see its value. At that time, early in 1913, Bohr was already convinced that part of the answer to the puzzle lay in introducing Planck's constant, h, into the equations describing the atom. Rutherford's atom only had two kinds of fundamental numbers incorporated in its structure, the charge on the

---

*A simple version of the formula says that the wavelengths of the first four hydrogen lines are given by multiplying a constant (36.456×10^{-5}) by 9/5, 16/12, 25/21, and 36/32. In this version of the formula, the top of each fraction is given by the sequence of squares (3^2, 4^2, 5^2, 6^2; the denominators are differences of squares, $3^2 - 2^2$, $4^2 - 2^2$, and so on.

electron, e, and the masses of the particles involved. No matter how much you juggle with the figures, you can't get a number that has the dimensions of length out of a mixture of mass and charge, so that Rutherford's model had no "natural" unit of size. But with an action, like h, added to the brew it is possible to construct a number that has the dimensions of length and can be regarded, in a rough and ready sort of way, as revealing something about the size of the atom. The expression h^2/me^2 is numerically equivalent to a length, about 20×10^{-8} cm, which is very much in the ball park required to fit in with the properties of atoms inferred from the scattering experiments and other studies. To Bohr, it was clear that h belonged in the theory of atoms. The Balmer series showed him just where it belonged.

How can an atom produce a very sharp spectral line? By either emitting or absorbing energy with a very precise frequency, v. Energy is related to frequency by Planck's constant ($E = hv$), and if an electron in an atom emits a quantum of energy hv then the energy of the electron must change by precisely the corresponding amount E. Bohr said that the electrons "in orbit" around the nucleus of an atom stayed in place because they could not radiate energy continuously, but they would, on this picture, be allowed to radiate (or absorb) a whole quantum of energy—one photon—and jump from one energy level (one orbit, on the old picture) to another. This seemingly simple idea actually marks another profound break with classical ideas. It's as if Mars disappeared from its orbit and reappeared in the orbit of the Earth, instantaneously, while radiating off into space a pulse of energy (in this case, it would be gravitational radiation). You can see at once how poor the idea of a solar system atom is at explaining what goes on, and how much better it is to think of the electrons simply as in different states, corresponding to different energy levels, inside the atom.

A jump from one state to another can occur in either direction, up or down the ladder of energy. If an atom absorbs light, then the quantum hv is used to move the electron up an energy level (to a higher rung on the ladder); if the electron then falls back to its original state precisely the same energy hv will be radiated. The mysterious constant 36.456×10^{-5} in Balmer's formula could be written naturally in terms of Planck's constant, and that meant Bohr could calculate the possible energy levels "allowed" for the single electron in the hydrogen atom, and the measured frequency of the spectral lines could now be interpreted as revealing how much energy difference there was between the different levels.*

HYDROGEN EXPLAINED

After discussing his work with Rutherford, Bohr published his theory of the atom in a series of papers during 1913. The theory worked very well for hydrogen, and looked as if it might be capable of being developed to account for the spectra of more complicated atoms as well. In September, Bohr attended the eighty-third annual meeting of the British Association for the Advancement of Science, and described his work to an audience that included many of the most eminent atomic physicists of the day. By and large, his report was well received, and Sir James Jeans called it ingenious, suggestive, and convincing. J. J. Thomson was among those who remained unconvinced, but thanks to this meeting even scientists who weren't persuaded by the arguments had at least heard of Bohr and his work on atoms.

---

*When dealing with electrons and atoms everyday energy units are rather too large for convenience, and the appropriate unit is the electron Volt (eV), which is the amount of energy an electron would pick up in moving across an electric potential difference of one Volt. The unit was introduced in 1912. In more everyday terms, an electron Volt is 1.602×10^{-19} joule, and one Watt is one joule per second. An ordinary light bulb burns energy at a rate of 100 Watts, which if you want to you can express as 6.24×10^{20} eV per second. It certainly sounds more impressive to say my light radiates six and a quarter hundred million trillion electron volts a second, but the energy is the same as when it was a hundred-watt lamp. The energies involved in electron transitions that produce spectral lines are a few eV—it takes just 13.6 eV to knock the electron right out of a hydrogen atom. The energies of particles produced by radioactive processes are several millions of electron Volts, or MeV.

Thirteen years after Planck's desperate measure of incorporating the quantum into the theory of light, Bohr introduced the quantum into the theory of the atom. But it
was to be another thirteen years before a true quantum theory emerged. In that time, progress was painfully slow—
one step backward for every two steps forward, and sometimes two steps backward for every one that had seemed to
be going in the right direction. Bohr's atom was a
hodgepodge. It mixed quantum ideas with those of classical
physics, using whatever mixture seemed necessary to
patch things up and keep the model going. It "allowed"
many more spectral lines than can actually be seen in the
light from different atoms, and arbitrary rules had to be
brought in to say that some transitions between different
energy states within the atom were "forbidden." New properties of the atom—quantum numbers—were assigned ad
hoc to fit the observations, with no underpinning of a secure theoretical foundation to explain why these quantum
numbers were required, or why some transitions were forbidden. In the middle of all this, the European world was
disrupted by the outbreak of the First World War, the year
after Bohr introduced his first model of the atom.

As in every other sphere of life, science would never be
the same again after 1914. The war stopped the easy movement of researchers from one country to another, and ever
since the First World War some scientists in some countries
have found it difficult to communicate with all of their colleagues around the world. The war also had a direct effect
on scientific research in the great centers where physics
had been making so much progress in the early years of the
twentieth century. In the belligerent nations, the young
men left the laboratories and went off to war, leaving the
older professors, such as Rutherford, to carry on as best
they could; many of those young men, the generation that
should have picked up Bohr's ideas and taken them on in
the years after 1913, died in action. The work of neutral
scientists was also affected, although in some ways some of
them may have benefited from the misfortunes of others.
Bohr himself was appointed Reader in Physics in Manchester; in Göttingen a Dutch citizen, Peter Debye, carried

out important studies of the structure of crystals, using X rays as probes. Holland and Denmark, indeed, remained scientific oases at that time, and Bohr returned to Denmark in 1916 to become Professor of Theoretical Physics in Copenhagen, and then to found, in 1920, the research institute that bears his name. News from a German researcher such as Arnold Sommerfeld (one of the physicists who refined Bohr's model of the atom, to such an extent that the model was sometimes referred to as the "Bohr-Sommerfeld" atom) could pass to neutral Denmark, and then from Bohr on to Rutherford in England. Progress continued to be made, but it was not the same.

After the war, German and Austrian scientists were not invited to international conferences for many years; Russia was in a revolutionary turmoil; science had lost some of its internationalism as well as a generation of young men. It fell to a completely new generation to take quantum theory from the halfway house of Bohr's hodgepodge atom (which had, admittedly, been refined by the diligent efforts of many researchers into a remarkably effective, if ramshackle, contrivance) on to its full glory as quantum mechanics. The names of that generation resound through modern physics—Werner Heisenberg, Paul Dirac, Wolfgang Pauli, Pascual Jordan, and others. They were members of the first quantum generation, born and bred in the years after Planck's great contribution (Pauli in 1900, Heisenberg in 1901, Dirac and Jordan in 1902), and coming into scientific research in the 1920s. They had no ingrained training in classical physics to overcome, and less need than even so brilliant a scientist as Bohr of half-measures to retain a flavor of classical ideas in their theories of the atom. It is entirely appropriate, and perhaps no coincidence, that the time from Planck's discovery of the blackbody equation to the flowering of quantum mechanics was just twenty-six years, the time it took for a generation of new physicists to develop into research scientists. That generation, however, had two great legacies from its still active elders, apart from Planck's constant itself. The first was Bohr's atom, providing a clear indication that quantum ideas *had* to be incorporated into any satisfactory theory of atomic processes; the second came from the one great scientist of the time who never had seemed hamstrung by the ideas of classical physics, the exception to all the rules. In 1916, at the peak of the war and working in Germany, Einstein introduced the notion of probability into atomic theory. He did so as an expedient—another contribution to the hodgepodge that made the workings of Bohr's atom resemble the observed behavior of real atoms. But this expedient outlasted the Bohr atom to become the underpinning foundation of the true quantum theory—even though, ironically, it was later disowned by Einstein himself in his famous comment, "God does not play dice."

AN ELEMENT OF CHANCE: GOD'S DICE

Back in the early 1900s, when Rutherford and his colleague Frederick Soddy were investigating the nature of radioactivity, they had discovered a curious and fundamental property of the atom, or rather of its nucleus. Radioactive "decay," as it became known, had to involve a fundamental change in an individual atom (we now know that it involves the breakup of a nucleus, and the ejection of parts of the nucleus), but it seemed to be unaffected by any outside influence. Heat the atoms up or cool them down, put them in a vacuum or a bucket of water, the process of radioactive decay proceeds undisturbed. There seemed to be no way of predicting exactly when a particular atom of a radioactive substance would decay, emitting an alpha or beta particle and gamma rays, but the experiments showed that out of a large number of radioactive atoms of the same element a certain proportion would always decay in a certain time. In particular, for every radioactive element there is a characteristic time called the half-life, during which exactly half of the atoms in a sample decay. Radium, for example, has a half-life of 1,600 years; a radioactive form of carbon, called carbon-14, has a half-life of a little under 6,000 years, which makes it useful in archaeological dating; and radioactive potassium decays with a half-life of 1,300 million years.

Without knowing what made one atom in a vast array of atoms disintegrate while its neighbors did not, Rutherford and Soddy used this discovery as the basis of a statistical theory of radioactive decay, a theory using actuarial techniques like those used by insurance companies, who know that though some of the people they insure will die young and their heirs will receive far more from the insurance company than the premiums paid in, other customers will live long and pay enough in premiums to compensate. Without knowing which clients will die when, the actuarial tables enable the accountants to balance the books. In the same way, statistical tables allow physicists to balance the books of radioactive decay, provided they are dealing with large collections of atoms.

One curious feature of this behavior is that radioactivity never quite disappears from a sample of radioactive material. From the millions of atoms present, half decay in a certain time. Over the next half-life—exactly the same time span—half of the rest decay, and so on. The number of radioactive atoms left in the sample gets smaller and smaller, closer and closer to zero, but each step toward zero takes it only half the way there.

In those early days, physicists such as Rutherford and Soddy imagined that eventually someone would find out exactly what made an individual atom decay, and that this discovery would explain the statistical nature of the process. When Einstein took the statistical techniques over into the Bohr model to account for details of atomic spectra, he too anticipated that later discoveries would remove the need for the "actuarial tables." They were all wrong.

The energy levels of an atom, or of an electron in an atom, can be thought of as a flight of steps. The depth of each of the steps is not equal in terms of energy—the top steps are closer together than the bottom ones. Bohr showed that in the case of hydrogen, the simplest atom, the energy levels could be represented in terms of a staircase where the depth of each tread below the top of the staircase

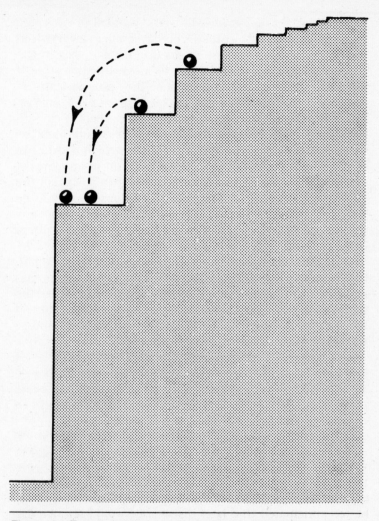

Figure 4.1/Energy levels in a simple atom such as
hydrogen can be compared with a set of steps having
different depths. A ball placed on different steps
represents an electron in different energy levels in the
atom. Moving down from one step to another corresponds
to the release of a precise amount of energy, responsible
in the hydrogen atom for the lines of the Balmer series
in the spectrum. There are no in-between lines because
there are no in-between "steps" for the electron to "rest" on.

is proportional to $1/n^2$, where n is the number of each step from the bottom. A transition from level one to level two on this staircase requires that an electron takes in exactly the amount of energy $h\nu$ required to move up that step; if the electron falls back to level one (the "ground state" of the atom) then it gives up the same amount of energy. There is no way in which a ground-state electron can absorb less energy, because there is no in-between "step" on which it can rest, and there is no way in which a level-two electron can emit less than this quantum of energy, because there is nowhere for it to jump down to except the ground state. Because there are many steps on which the electron can lodge, and because it can jump or fall from any step to any other, there are many lines in the spectrum of each element. Each line corresponds to a transition between steps—between energy levels with different quantum numbers. All the transitions that end up in the ground state, for example, produce a family of spectral lines like the Balmer series; all the transitions from higher steps that end up on level two correspond to another set of lines, and so on.* In a hot gas atoms are constantly colliding with one another, so that electrons are excited to high energy levels and then fall back, radiating bright spectral lines as they do so. When light passes through a cold gas, the ground-state electrons are raised to higher energies, absorbing light as they do so and leaving dark lines in the spectrum.

If the Bohr model of the atom meant anything at all, this explanation of how hot atoms radiate energy ought to tie in with Planck's law. The blackbody spectrum of cavity radiation ought to be simply the combined effect of a lot of atoms radiating energy as electrons jump from one energy level to another.

In 1916 Einstein had completed his General Theory of Relativity, and he turned his attention once again to quantum theory (compared with his masterwork, this may have seemed like recreation). He was probably encouraged by the success of Bohr's model of the atom, and also at this

---

*In fact, the Balmer series in hydrogen's spectrum does correspond to the transitions that end on level two.

time his own version of the corpuscular theory of light at last began to gain ground. Robert Andrews Millikan, an American physicist, had been one of the strongest opponents of Einstein's interpretation of the photoelectric effect, when this interpretation first appeared back in 1905. He spent ten years testing the idea in a series of superb experiments, starting out with the aim of proving Einstein wrong and ending up in 1914 with direct experimental proof that Einstein's explanation of the photoelectric effect in terms of light quanta, or photons, was correct. In the process, he derived a very accurate experimental determination of the value of h, and in 1923, to complete the irony, he received the Nobel Prize for this work and his measurement of the charge on the electron.

Einstein realized that the decay of an atom from an "excited" energy state—with an electron at a high energy level—into a state with less energy—with the electron at a lower energy level—is very similar to the radioactive decay of an atom. He used the statistical techniques developed by Boltzmann (for dealing with the behavior of collections of atoms) to deal with individual energy states, working out the probability that a particular atom would be in an energy state corresponding to a particular quantum number n, and he used the probabilistic "actuarial tables" of radioactivity to work out the likelihood that an atom in state n will "decay" into another energy state with less energy (that is, with a lower quantum number). This all led, in a clear and simple way, to Planck's formula for blackbody radiation, derived entirely on the basis of quantum ideas. Soon, using Einstein's statistical ideas, Bohr was able to extend his model of the atom, taking on board the explanation that some lines in the spectrum are more pronounced than others because some transitions between energy states are more probable—more likely to happen—than others. He could not explain *why* this should be so, but nobody was too worried about that at the time.

Like the people who studied radioactivity in those days, Einstein believed that the actuarial tables were not the last word, and that later research would determine *why* a particular transition occurred exactly when it did, and not at some other time. But it was at this point that quantum theory really began to cut loose from classical ideas, and no "underlying reason" for radioactive decay or atomic-energy transitions to occur when they do has ever been found. It really does seem that these changes occur entirely by chance, on a statistical basis, and that already begins to raise fundamental philosophical questions.

In the classical world, everything has its cause. You can trace the cause of any event backward in time to find the cause of the cause, and what caused that, and so on back to the Big Bang (if you are a cosmologist) or to the moment of Creation in a religious context, if that is the model you subscribe to. But in the world of the quantum, such direct causality begins to disappear as soon as we look at radioactive decay and atomic transitions. An electron doesn't move down from one energy level to another at a particular time for any particular reason. The lower energy level is more desirable for the atom, in a statistical sense, and so it is quite likely (the amount of likelihood can even be quantified) that the electron will make such a move, sooner or later. But there is no way to tell when the transition will occur. No outside agency pushes the electron, and no internal clockwork times the jump. It just happens, for no particular reason, now rather than then.

This isn't much of a break with strict causality, and although many nineteenth-century scientists might have been horrified by the idea I doubt if any of the readers of this book are too concerned. But it is merely the tip of an iceberg, the first clue to the real strangeness of the quantum world, and worth noting even though its significance was not appreciated at the time. It came in 1916, and it came from Einstein.

ATOMS IN PERSPECTIVE

It would be tedious to elaborate all the detailed refinements that went into Bohr's model of the atom in the years up to

1926, and even more tedious to reveal only then that most of this groping toward the truth was wrong anyway. But the Bohr atom has such a grip on the textbooks and popularizations that it cannot be ignored, and in its final form it does represent just about the last model of the atom that bears any relation to the images we are used to in everyday life. The indivisible billiard ball atom of the ancients has been shown to be not just divisible but mostly empty space, full of strange particles doing strange things. Bohr provided a framework that puts some of those strange things in a context similar to everyday life; and although in some ways it might be better to discard all everyday ideas before plunging fully into the world of the quantum, most people seem happier to pause and survey the Bohr model before taking the plunge. Halfway between classical physics and quantum theory, let's pause for breath and rest awhile before we move on into unknown territory. But let's not waste time and energy tracing all the mistakes and half-truths involved in the patchwork development of the Bohr model and the nucleus in the years up to 1926. Instead, I will use the perspective of the 1980s to look back at Bohr's atom and to describe a kind of modern synthesis of Bohr's ideas, and those of his colleagues, including some pieces of the puzzle that were, in fact, only fitted into place much later on.

Atoms are very small. Avogadro's Number is the number of atoms of hydrogen in one gram of the gas. Hydrogen gas isn't the sort of thing we meet up with in everyday life, however, so to get some idea of just how small atoms are let's think instead of a lump of carbon—coal, diamond, or soot. Because each atom of carbon weighs twelve times as much as an atom of hydrogen, the same number of carbon atoms as in a gram of hydrogen weighs twelve grams. Ten grams weigh a little over a third of an ounce, twelve grams is just under half an ounce. A spoonful of sugar, a rather large diamond or a rather small lump of coal each weigh about half an ounce. And that is how much carbon contains Avogadro's Number of atoms, 6×10^{23} (a 6 followed by 23 zeroes) atoms. How can we put that number in perspective? Huge numbers are often called "astronomical," and many astronomical numbers are indeed big, so let's try to find a comparably big number in astronomy.

The age of the universe, astronomers believe, is roughly 15 billion years, 15×10^9 yr. Clearly, 10^{23} is a lot bigger than 10^9. Let's turn the age of the universe into an even bigger number, using the smallest unit of time with which we might feel familiar, one second. Each year contains 365 days, each day 24 hours, each hour 3,600 seconds. In round terms, each year contains 32 million seconds, about 3×10^7 sec. So 15 billion years contains 45×10^{16} seconds, following the rule that you multiply numbers like 10^9 and 10^7 by adding the exponents to give 10^{16}. So, again in round terms, the age of the universe in seconds is 5×10^{17}.

That is still a long way short of 6×10^{23}—six powers of ten short. That doesn't look too bad when there are 23 powers of ten to play with, but what does it really mean? We divide 6×10^{23} by 5×10^{17} and, subtracting the exponents, we get a bit more than 1×10^6—a million. Imagine a supernatural being watching our universe develop from the Big Bang of creation. This being is equipped with half an ounce of pure carbon, and a pair of tweezers so fine that it can pick out individual carbon atoms from the heap. Starting with the instant of the beginning of the Big Bang in which our universe was born, the being removes one carbon atom from the heap every second, and throws it away. By now, 5×10^{17} atoms will have been discarded; what proportion will remain? After all that activity, working steadily for 15 billion years, the supernatural being will have removed just about *one millionth* of the carbon atoms; what remains in the heap is still a million times more than has been discarded.

Now, perhaps, you have some idea how small an atom is. The surprise is not that Bohr's model of the atom is a rough and ready approximation, or that the rules of everyday physics don't apply to atoms. The miracle is that we understand anything at all about atoms, and that we *can* find ways to bridge the gap from classical Newtonian physics to atomic quantum physics.

As far as it is possible to build up a physical picture of anything so tiny, this is what an atom is like. As Rutherford showed, a tiny central nucleus is surrounded by a cloud of electrons, buzzing about it like bees. At first, it was thought that the nucleus consisted only of protons, each with a positive charge of the same size as an electron's negative charge, so that an equal number of protons and electrons made each atom electrically neutral; later on it turned out that there is another fundamental atomic particle that is very similar to the proton but has no electric charge. This is the neutron, and in all atoms except the simplest form of hydrogen there are neutrons as well as protons in the nucleus. But there are indeed the same number of protons as there are electrons in the neutral atom. The number of protons in the nucleus decides which element it is an atom of; the number of electrons in the cloud (the same as the number of protons) determines the chemistry of that atom, and that element. But because some atoms that have the same number of protons and electrons as each other may have a different number of neutrons, chemical elements can come in different varieties, called isotopes. The name was invented by Soddy in 1913, from the Greek for "same place," because of the discovery that atoms with different weights could belong in the same place in the table of chemical properties, the periodic table of the elements. In 1921 Soddy received the Nobel Prize (in chemistry) for his work on isotopes.

The simplest isotope of the simplest element is the most common form of hydrogen, in which one proton is accompanied by one electron. In deuterium, each atom consists of one proton *and* one neutron accompanied by one electron, but the chemistry is the same as ordinary hydrogen. Because neutrons and protons are very nearly the same mass as each other, and each is about 2,000 times as massive as an electron, the total number of protons plus neutrons in a nucleus determines all but a small fraction of the mass of an atom. This is usually denoted by the number A, called the mass number. The number of protons in the nucleus, which determines the properties of the element, is called the atomic number, Z. The unit in which atomic masses are measured is called, logically enough, the atomic mass unit, and it is defined as one twelfth of the mass of the isotope of carbon, which contains six neutrons and six protons in its nucleus. This isotope is called carbon-12, or in shorthand written as ^{12}C; other isotopes are ^{13}C and ^{14}C, which contain seven and eight neutrons per nucleus, respectively.

The more massive a nucleus is (the more protons it contains) the more variety of isotopes. Tin, for example, has fifty protons in its nucleus ($Z = 50$) and ten stable isotopes with mass numbers ranging from $A = 112$ (62 neutrons) to $A = 124$ (74 neutrons). There are always at least as many neutrons as protons in stable nuclei (except for the simplest hydrogen atom); the neutral neutrons help to hold the positive protons, which have a tendency to repel each other, together. Radioactivity is associated with unstable isotopes that change into a stable form and emit radiation as they do so. A beta ray is an electron ejected as a neutron turns into a proton; an alpha particle is an atomic nucleus in its own right, two protons and two neutrons (the nucleus of helium-4) ejected as an unstable nucleus adjusts its internal structure; and very massive unstable nuclei split into two or more lighter, stable nuclei through the now well-known process of nuclear, or atomic, fission, with alpha and beta particles also emerging from the brew. All of this goes on in a volume that is almost unimaginably smaller than the almost unimaginably small size of the atom itself. A typical atom is about 10^{-10} of a meter across; the nucleus about 10^{-15} m in radius, 10^5 times smaller than the atom. Because volumes go as the cube of radius, we have to multiply the exponent by three to find that the volume of the nucleus is 10^{15} times smaller than the volume of the atom.

CHEMISTRY EXPLAINED

The cloud of electrons provides the outward face of the atom and the means by which it interacts with other atoms. It is largely immaterial what lies buried that far into the heart of the electron cloud—what another atom "sees" and "feels" are the electrons themselves, and it is the interactions between the electron clouds that are responsible for chemistry. By explaining the broad features of the electron cloud, Bohr's model of the atom put chemistry onto a scientific footing. Chemists already knew that some elements were very alike in their chemical properties, even though they had different atomic weights. When the elements are arranged in a table according to their atomic weight (and especially when allowance is made for different isotopes) these similar elements show up at regular intervals, one pattern recurring for elements eight atomic numbers apart, for example. This gives the table, when arranged so that elements with similar properties are grouped together, its name "periodic."

In June 1922 Bohr visited the University of Göttingen in Germany, to give a series of lectures on quantum theory and atomic structure. Göttingen was about to become one of the three key centers in the development of the complete version of quantum mechanics, under the direction of Max Born, who became Professor of Theoretical Physics there in 1921. He had been born in 1882, son of the Professor of Anatomy at the University of Breslau, and was a student in the early 1900s, at the time Planck's ideas first appeared. At first he studied mathematics, only turning to physics (and working for a time at the Cavendish) after completing his doctorate in 1906. This, as we shall see, turned out to have been an ideal training in the years ahead. An expert on relativity, Born's work was always characterized by full mathematical rigor, in striking contrast to Bohr's patchwork theoretical edifices, built with the aid of brilliant insights and physical intuition, but often leaving others to catch up with the mathematical details. Both kinds of genius were essential to the new understanding of atoms.

Bohr's lectures in June 1922 were a major event in the renewal of German physics after the war, and also in the history of quantum theory. They were attended by scientists from all over Germany, and became known (with a not-too-subtle pun on certain other famous German celebrations) as the "Bohr Festival." And in those lectures, after carefully preparing his ground, Bohr presented the first successful

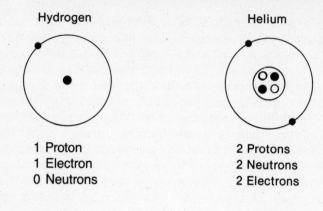

Hydrogen

1 Proton
1 Electron
0 Neutrons

Helium

2 Protons
2 Neutrons
2 Electrons

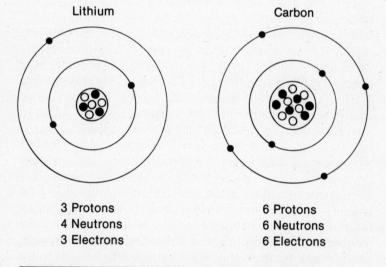

Lithium

3 Protons
4 Neutrons
3 Electrons

Carbon

6 Protons
6 Neutrons
6 Electrons

*Figure 4.2/*Atoms of some of the simplest elements can
be represented for many purposes as a nucleus surrounded
by electrons in shells corresponding to the steps on the
energy-level staircase. The quantum rules allow only two
electrons on the lowest step, so lithium, with three electrons,
has to put one of them onto the next step up the energy ladder.
This second shell has "room" for eight electrons, so that
carbon has a shell exactly half full, which is the reason for
its interesting chemical properties as the basis of life.

theory of the periodic table of the elements, a theory that
survives in essentially the same form to this day. Bohr's idea
stemmed from a picture of the electrons being added to the
nucleus of an atom. Whatever the atomic number of that
nucleus, the first electron would go into an energy state
corresponding to the ground state of hydrogen. The next
electron would go into a similar energy state, giving an outward appearance rather like the helium atom, which has
two electrons. But, said Bohr, there was no room for any
more electrons at that level in the atom, and the next one to
be added would have to go into a different kind of energy
level. So an atom with three protons in its nucleus and
three electrons outside the nucleus should have two of
those electrons more tightly tied to the nucleus and one left
over; it ought to behave rather like a one-electron atom (hydrogen) as far as chemistry is concerned. The $Z = 3$ element is lithium, and it does indeed show some chemical
similarities to hydrogen. The next element in the periodic
table with similar properties to lithium is sodium, with
$Z = 11$, eight places beyond lithium. So Bohr argued that
there must be eight places available in the set of energy
levels outside the inner two electrons, and that when these
were filled the next electron, the eleventh in all, had to go
into another energy state still less tightly tied to the nucleus, again mimicking the appearance of an atom with
only one electron.

These energy states are called "shells," and Bohr's explanation of the periodic table involved successively filling
up the shells with electrons as Z increased. You can think
of the shells as onion skins wrapped around one another;
what matters for chemistry is the number of electrons in
the outermost shell of the atom. What goes on deeper inside
plays only a secondary role in determining how the atom
will interact with other atoms.

Working outward through the electron shells, and incorporating all the evidence from spectroscopy, Bohr explained the relationships between the elements in the
periodic table in terms of atomic structure. He had no idea
why a shell containing eight electrons should be full
("closed"), but he left none of his audience in any doubt

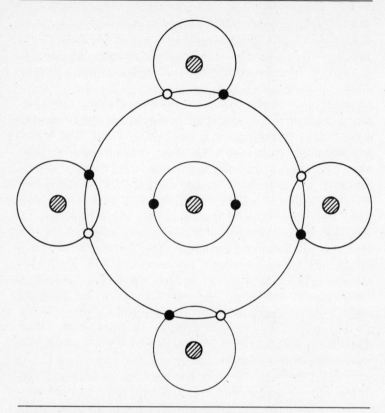

*Figure 4.3/*When one carbon atom links with four atoms of
hydrogen, electrons are shared in such a way that each
hydrogen atom has the illusion of a full innermost shell (two
electrons) and each carbon atom "sees" eight electrons in
its second shell. This is a very stable configuration.

that he had discovered the essential truth. As Heisenberg
said later, Bohr "had not proved anything mathematically . . . he just knew that this was more or less the
connection."* And Einstein commented in his *Autobiographical Notes* in 1949, describing the success of
Bohr's work based on quantum theory, "that this insecure
and contradictory foundation was sufficient to enable a

*Quoted in Mehra and Rechenberg, volume 1 page 357.

man of Bohr's unique instinct and tact to discover the major laws of spectral lines and of the electron-shells of the atoms together with their significance for chemistry appeared to me like a miracle—and appears to me as a miracle even today.*

Chemistry is concerned with the way atoms react and combine to make molecules. Why does carbon react with hydrogen in such a way that four atoms of hydrogen attach to one of carbon to make one molecule of methane? Why does hydrogen come in the form of molecules, each made of two atoms, while helium atoms do not form molecules? And so on. The answers came with stunning simplicity from the shell model. Each hydrogen atom has one electron, whereas helium has two. The "innermost" shell would be full if it had two electrons in it, and (for some unknown reason) filled shells are more stable—atoms "like" to have filled shells. When two hydrogen atoms get together to form a molecule, they share their two electrons in such a way that each feels the benefit of a closed shell. Helium, having a full shell already, is not interested in any such proposition and disdains to react chemically with anything.

Carbon has six protons in its nucleus and six electrons outside. Two of these are in the inner closed shell, leaving four associated with the next shell, which is half empty. Four hydrogen atoms can each claim a part share in one of the four outer carbon electrons and contribute their own electron to the deal. Each hydrogen atom ends up with a pseudoclosed shell of two inner electrons, while each carbon atom has a pseudoclosed second shell of eight electrons.

Atoms combine, said Bohr, in such a way that they get as close as they can to making a closed outer shell. Sometimes, as with the hydrogen molecule, it is best to think of a pair of electrons being shared by two nuclei; in other cases, an appropriate picture is to imagine an atom that has an odd electron in its outer shell (sodium, perhaps) giving the electron away to an atom that has an outer shell containing seven electrons and one vacancy (in this case, it might be

---

*Op. cit., page 359.

chlorine). Each atom is happy—the sodium, by losing an electron, leaves a deeper, but filled, shell "visible"; the chlorine, by gaining an electron, fills its outermost shell. The net result, however, is that the sodium atom has become a positively charged ion by losing one unit of negative charge, while the chlorine atom has become a negative ion. Since opposite charges attract, the two stick together to form an electrically neutral molecule of sodium chloride, common salt.

All chemical reactions can be explained in this way, as a sharing or swapping of electrons between atoms in a bid to achieve the stability of filled electron shells. Energy transitions involving outer electrons produce the characteristic spectral fingerprint of an element, but energy transitions involving deeper shells (and therefore much more energy, in the X-ray part of the spectrum) should be the same for all elements, as indeed they prove to be. Like all the best theories, Bohr's model was confirmed by a successful predic-

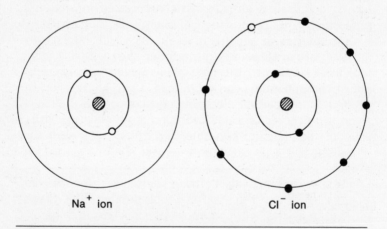

Na⁺ ion   Cl⁻ ion

Figure 4.4/By giving up its lone outer electron, a sodium atom achieves a desirable quantum mechanical configuration and is left with a positive charge. By accepting an extra electron, chlorine fills its outer shell with eight electrons and gains a negative charge. The charged ions are then held together to make molecules and crystals of common salt (NaCl) by electrostatic forces.

tion. With the elements arranged in a periodic table, even in 1922 there were a few gaps, corresponding to undiscovered elements with atomic numbers 43, 61, 72, 75, 85 and 87. Bohr's model predicted the detailed properties of these "missing" elements and suggested that element 72, in particular, should have properties similar to zirconium, a forecast that contradicted predictions made on the basis of alternative models of the atom. The prediction was confirmed within a year with the discovery of hafnium, element 72, which turned out to have spectral properties exactly in line with those predicted by Bohr.

This was the high point of the old quantum theory. Within three years, it had been swept away, although as far as chemistry is concerned you need little more than the idea of electrons as tiny particles orbiting around atomic nuclei in shells that would "like" to be full (or empty, but preferably not in between).* And if you are interested in the physics of gases, you need little more than the image of atoms as hard, indestructible billiard balls. Nineteenth-century physics will do for everyday purposes; the physics of 1923 will do for most of chemistry; and the physics of the 1930s takes us about as far as anyone has yet gone in the search for ultimate truths. There has been no great breakthrough comparable to the quantum revolution for fifty years, and in all that time the rest of science has been catching up with the insights of a handful of geniuses. The success of the Aspect experiment in Paris in the early 1980s marked the end of that catching-up period, with the

---

*I am, of course, exaggerating the simplicity of chemistry here. The "little more" that is needed to explain more complex molecules was developed in the late 1920s and early 1930s, using the fruits of the full development of quantum mechanics. The person who did most of the work was Linus Pauling, more familiar today as a peace campaigner and proponent of vitamin C, who received the first of his two Nobel Prizes for the work, being cited in 1954 "for his research into the nature of the chemical bond and its application to the elucidation of the structure of complex substances." Those "complex substances" elucidated with the aid of quantum theory by Pauling, a physical chemist, opened the way to a study of the molecules of life. The key significance of quantum chemistry to molecular biology is acknowledged by Horace Judson in his epic book *The Eighth Day of Creation;* the detailed story, alas, is beyond the scope of the present book.

first direct experimental proof that even the most strange aspects of quantum mechanics are a literal description of the way things are in the real world. The time has come to discover just how strange the world of the quantum really is.

PART TWO

QUANTUM MECHANICS

"All science is either
physics or stamp collecting."

ERNEST RUTHERFORD
1871–1937

CHAPTER FIVE

PHOTONS AND ELECTRONS

In spite of the success of Planck and Bohr in pointing the way toward a physics of the very small that differed from classical mechanics, quantum theory as we know it today only really began with the acceptance of Einstein's idea of the light quantum, and the realization that light had to be described *both* in terms of particles *and* waves. And even though Einstein first introduced the light quantum in his 1905 paper on the photoelectric effect, it was not until 1923 that the idea became accepted and respectable. Einstein himself moved cautiously, well aware of the revolutionary implications of his work, and in 1911 he told the participants at the first Solvay Congress, "I insist on the provisional character of this concept, which does not seem reconcilable with the experimentally verified consequences of the wave theory."*

---

*The Solvay Congresses were a series of scientific meetings sponsored by Ernest Solvay, a Belgian chemist who made a fortune from his method for manufacturing sodium carbonate. Because of his interest in more abstract science, Solvay provided funds for these meetings at which the leading physicists of the day were able to meet and exchange views.

Although Millikan proved by 1915 that Einstein's equation for the photoelectric effect was correct, it still seemed unreasonable to accept the reality of particles of light, and looking back on his work from the 1940s Millikan commented on his tests of this equation, "I was compelled in 1915 to assert its unambiguous verification in spite of its unreasonableness . . . it seemed to violate everything we knew about the interference of light." At the time, he expressed himself more forcefully. Reporting the experimental verification of the accuracy of Einstein's equation of the photoelectric effect, he went on to say, "The semicorpuscular theory by which Einstein arrived at his equation seems at present wholly untenable." This was written in 1915; in 1918 Rutherford commented that there seemed to be "no physical explanation" for the connection between energy and frequency that Einstein had explained thirteen years before with his hypothesis of light quanta. It wasn't that Rutherford didn't know of Einstein's suggestion, but that he wasn't convinced by it. Since all the experiments designed to test the wave theory of light showed light to be made up of waves, how could light be made of particles?*

PARTICLES OF LIGHT

In 1909, about the time that he ceased being a patent clerk and took up his first academic post, as an associate professor in Zürich, Einstein made a small but significant step forward, referring for the first time to "pointlike quanta with energy $h\nu$". Particles like electrons are represented by "pointlike" objects in classical mechanics, and this is a far cry from any description in terms of waves, except that the frequency of the radiation, ν, tells us the energy of the particle. "It is my opinion," said Einstein in 1909, "that the next phase in the development of theoretical physics will bring us a theory of light that can be interpreted as a kind of fusion of the wave and the emission theory."

*The quotes in this passage are taken from A. Pais's *Subtle Is the Lord*.

This comment, scarcely noticed at the time, strikes to the heart of modern quantum theory. In the 1920s, Bohr expressed this new basis of physics as the "principle of complementarity," which holds that the wave and particle theories of (in this case) light are not mutually exclusive to one another but complementary. *Both* concepts are necessary to provide a complete description, and this shows up strikingly in the need to measure the energy of the light "particle" in terms of its frequency, or wavelength.

Soon after he made these remarks, however, Einstein left off serious thinking about quantum theory while he developed his General Theory of Relativity. When he returned to the quantum fray in 1916, it was with another logical development on the light-quantum theme. His statistical ideas helped, as we have seen, to tidy up the picture of the Bohr atom and improved Planck's description of blackbody radiation. These calculations of the way matter absorbs or emits radiation also explained how momentum is transferred from radiation to matter, provided that each quantum of radiation $h\nu$ carries with it a momentum $h\nu/c$. The work harks back to another of the great 1905 papers, on Brownian motion. Just as pollen grains are buffeted by the atoms of a gas or liquid, so that their motion proves the reality of the atoms, so the atoms are themselves buffeted by the "particles" of blackbody radiation. This "Brownian motion" of atoms and molecules could not be observed directly, but the buffeting causes statistical effects that could be measured in terms of properties such as the pressure of a gas. It was the statistical effects that Einstein explained in terms of blackbody radiation particles which carry momentum.

However, the same expression for the momentum of a "particle" of light comes straight out of special relativity, in a very simple way. In relativity theory, the energy (E), momentum (p) and rest mass (m) of a particle are related by the simple equation

$$E^2 = m^2c^4 + p^2c^2$$

Since the particle of light has no rest mass, this equation very quickly reduces to

$$E^2 = p^2c^2$$

or simply $p = E/c$. It may seem surprising that it took Einstein so long to spot this, but then he had had other things—like general relativity—on his mind. Once he did make the connection, however, the agreement between the statistical arguments and relativity theory certainly made the case a lot stronger. (From another point of view, since the statistics show that $p = E/c$, you can argue that the relativistic equations then establish that the particle of light has zero rest mass.)

It was this work that convinced Einstein himself that light quanta were real. The name "photon" for the particle of light was not introduced until 1926 (by Gilbert Lewis, based in Berkeley, California), and it only became part of the language of science after the fifth Solvay Congress was held under the title "Electrons and Photons" in 1927. But although in 1917 Einstein stood alone in his belief in the reality of what we now call photons, this seems an appropriate time to introduce the name. It was another six years before incontrovertible, direct experimental proof of the reality of photons was obtained, by the American physicist Arthur Compton.

Compton had been working with X rays since 1913. He worked in several American universities and at the Cavendish in England. Through a series of experiments in the early 1920s he was led inexorably to the conclusion that the interaction between X rays and electrons could only be explained if the X rays were treated in some ways as particles—photons. The key experiments concern the way in which the X-radiation is scattered by an electron—or, in particle language, the way a photon and an electron interact when they collide. When an X-ray photon hits an electron, the electron gains energy and momentum and moves off at an angle. The photon itself loses energy and momentum and moves off at a different angle, which can be calculated from the simple laws of particle physics. The collision is like the impact of a moving billiard ball on a stationary ball, and the transfer of momentum occurs in just the same way. In the case of the photon, however, the loss of energy means a change in frequency of the radiation, by the amount $h\nu$ given up to the electron. You need both descriptions, particle and wave, to get a complete explanation of the experiment. When Compton made the experiments, he found the interaction behaving exactly in accordance with this description—the scattering angles, the wavelength changes and the recoil of the electron all fitted perfectly with the idea that X-radiation comes in the form of particles with energy hv. The process is now called the Compton effect, and in 1927 Compton received a Nobel Prize for this work.* After 1923, the reality of photons as particles carrying both energy and momentum was established (although Bohr struggled hard for a time to find an alternative explanation of the Compton effect; he did not immediately see the need to include *both* particle and wave descriptions in a good theory of light, and saw the particle theory as a rival to the wave theory incorporated in his model of the atom). But all the evidence for the wave nature of light remained. As Einstein said in 1924, "there are therefore now two theories of light, both indispensable . . . without any logical connection."

The connection between those two theories formed the basis of the development of quantum mechanics in the next few hectic years. Progress was made on many different fronts simultaneously, and new ideas and new discoveries did not come neatly in the order they were required to build up the new physics. To tell a coherent story, I have to make the account more orderly than science itself was at the time, and one way of doing this is to lay the groundwork of relevant concepts before describing quantum mechanics itself, even though the theory of quantum mechanics began to be developed before some of those concepts were understood. Even the full implications of the particle/wave duality were not appreciated when quantum mechanics began to take shape—but in any logical description of quantum theory the next step after the discovery of the dual nature of light must be the discovery of the dual nature of matter.

---

*The theorist Peter Debye calculated the "Compton effect" independently at about the same time, and published a paper suggesting an experiment to test the idea. By the time his paper was published, Compton had already done the experiment.

PARTICLE/WAVE DUALITY

The discovery stemmed from a suggestion made by a French nobleman, Louis de Broglie. It sounds so simple, yet it struck to the heart of the matter. "If light waves also behave like particles," we can imagine de Broglie musing, "why shouldn't electrons also behave like waves?" If he had stopped there, of course, he would not have been remembered as one of the founders of quantum theory, nor would he have received a Nobel Prize in 1929. As an idle speculation the idea doesn't amount to much, and similar speculations had been aired about X rays long before Compton's work, at least as early as 1912, when the great physicist (and yet another Nobel Laureate) W. H. Bragg said of the state of X-ray physics at the time, "The problem becomes, it seems to me, not to decide between two theories of X rays, but to find . . . one theory which possesses the capacity of both."* De Broglie's great achievement was to take the idea of particle/wave duality and to carry it through mathematically, describing how matter waves ought to behave and suggesting ways in which they might be observed. He had one great advantage as a relatively junior member of the theoretical physics community, an elder brother, Maurice, who was a respected experimental physicist and who steered him toward the discovery. Louis de Broglie said later that Maurice stressed to him in conversations the "importance and undeniable reality of the dual aspects of particle and wave." This was an idea whose time had come, and Louis de Broglie was lucky to be around at the time when a conceptually simple piece of intuition could transform theoretical physics. But he certainly made the most of his intuitive leap.

De Broglie had been born in 1892. Family tradition had him destined for a career in the civil service, but when he entered the University of Paris in 1910 he became fired

---

*Quotes from de Broglie's writings, and Bragg, are taken from Max Jammer, *The Conceptual Development of Quantum Mechanics*

with an interest in science, especially quantum mechanics, a world opened to him partly by his brother (seventeen years his senior) who had obtained his doctorate in 1908 and who as one of the scientific secretaries to the first Solvay Congress passed on news to Louis. But after a couple of years his study of physics was interrupted in 1913 by what should have been a short period of compulsory military service but lasted, because of the First World War, until 1919. Picking up the threads after the war, de Broglie returned to the study of quantum theory, and began working along the lines that were to lead to his discovery of the underlying unity of particle and wave theories. The breakthrough came in 1923, when he published three papers on the nature of light quanta in the French journal *Comptes Rendus*, and wrote an English summary of the work which appeared in the *Philosophical Magazine* in February 1924. These short contributions made no great impact, but de Broglie immediately set about ordering his ideas and presenting them in a more complete form for his doctoral thesis. His examination at the Sorbonne was held in November 1924, and the thesis was published early in 1925 in the *Annales de Physique*. It was in that form that the basis of his work became clear and sparked one of the major advances in physics during the 1920s.

In his thesis, de Broglie started out from the two equations that Einstein had derived for light quanta,

$$E = h\nu \; ; \; p = h\nu/c$$

In both those equations, properties that "belong" to particles (energy and momentum) appear on the left, and properties that "belong" to waves (frequency) appear on the right. He argued that the failure of experiments to settle once and for all whether light is wave or particle must therefore be because the two kinds of behavior are inextricably tangled—even to measure the particle property of momentum you have to know the wave property called frequency. Yet this duality did not apply only to photons. Electrons were thought at the time to be good, well-behaved particles, except for the curious way they occupied distinct energy levels inside atoms. But de Broglie realized that the fact that electrons only existed in "orbits" defined by whole numbers (integers) also looked in some ways like a wave property. "The only phenomena involving integers in Physics were those of interference and of normal modes of vibration," he wrote in his thesis. "This fact suggested to me the idea that electrons too could not be regarded simply as corpuscles, but that periodicity must be assigned to them."

"Normal modes of vibration" are simply the vibrations that make the notes of a violin string or a sound wave in an organ pipe. A tightly stretched string, for example, might vibrate in such a way that each end is fixed while the middle wiggles to and fro. Touch the center of the string, and each half will vibrate in the same way, with the center at rest—and this higher "mode" of vibration also corresponds to a higher note, a harmonic, of the untouched full string. In the first mode, the wavelength is twice as long as in the second, and higher modes of vibration, corresponding to successively higher notes, can fit the vibrating string provided always that the length of the string is a whole number of wavelengths (1, 2, 3, 4, and so on). Only some waves, with certain frequencies, fit the string.

This is, indeed, similar to the way electrons "fit" into atoms in states corresponding to quantum-energy levels 1, 2, 3, 4, and so on. Instead of a stretched straight string, imagine one bent into a circle, an "orbit" around an atom. A standing vibration wave can run happily around the string, provided that the length of the circumference is a whole number of wavelengths. For any wave that did not precisely "fit" the string in this way, the wave would be unstable and dissipate as it interfered with itself. The head of the snake must always catch hold of its tail, or the string, like the analogy, falls apart. Could this explain the quantization of energy states in the atom, with each one corresponding to a resonating electron wave of a particular frequency? Like so many of the analogies based on the Bohr atom—indeed, like all physical pictures of the atom—the image is far from the truth, but helped toward a better understanding of the world of the quantum.

ELECTRON WAVES

De Broglie thought of the waves as being *associated* with particles, and suggested that a particle such as a photon is in fact guided on its way by the associated wave to which it is tied. The result was a thorough mathematical description of the behavior of light, which incorporated the evidence from both wave and particle experiments. The examiners who studied de Broglie's thesis liked the math, but did not believe that the proposal for a similar wave associated with a particle like the electron had any physical meaning—they regarded it as just a quirk of the mathematics. De Broglie did not agree. When asked by one of the examiners if an experiment could be devised to detect the matter waves, he said that it should be possible to make the required observations by diffracting a beam of electrons from a crystal. The experiment is just like the diffraction of light through not just two but an array of slits, with the gaps between the regularly spaced atoms in the crystal providing an array of "slits" narrow enough to diffract the high-frequency (small wavelength, compared with light or even X rays) electron waves.

De Broglie knew the right wavelength to look for, since by combining Einstein's two equations for light particles he obtained the very simple relation $p = h\nu/c$, which we've already met. Since wavelength is related to frequency by $\lambda = c/\nu$, this means $p\lambda = h$, or in plain words momentum multiplied by wavelength gives Planck's constant. The smaller the wavelength, the bigger the momentum of the corresponding particle, making electrons, with their small mass and correspondingly small momentum, the most "wavelike" of the particles then known. Just as in the case of light, or waves on the surface of the sea, diffraction effects only show up if the wave passes through a hole much smaller than its wavelength, and for electron waves that means a very small hole indeed, about the size of the gap between atoms in a crystal.

What de Broglie did not know was that effects that can

best be explained in terms of diffraction of electrons had been observed when beams of electrons were used to probe crystals as long ago as 1914. Two American physicists, Clinton Davisson and his colleague Charles Kunsman, had, indeed, been studying this peculiar behavior of electrons, scattering from crystals, during 1922 and 1923 while de Broglie was formulating his ideas. Ignorant of all this, de Broglie tried to persuade experimenters to carry out a test of the electron-wave hypothesis. Meanwhile, de Broglie's thesis supervisor, Paul Langevin, had sent a copy of the work to Einstein, who, hardly surprisingly, saw it as much more than a mathematical trick or analogy, and realized that matter waves must be real. He in turn passed the news on to Max Born in Göttingen, where the head of the experimental physics department, James Franck, commented that Davisson's experiments "had already established the existence of the expected effect"!*

Davisson and Kunsman, like other physicists, had thought the scattering effect to be caused by structure in the atoms being bombarded by electrons, not due to the nature of the electrons themselves. Walter Elsasser, a student of Born's, published a short note explaining the results of these experiments in terms of electron waves in 1925, but the experimenters were not impressed by this reinterpretation of their data by a theorist—especially not by an unknown twenty-one-year-old student. Even in 1925, in spite of the existing experimental evidence, the idea of matter waves remained no more than a vague notion. It was only when Erwin Schrödinger came up with a new theory of atomic structure incorporating de Broglie's idea but going far beyond it that the experimenters felt an urgent need to check the electron-wave hypothesis by performing the diffraction experiments. When the work was done, in 1927, it proved de Broglie to have been entirely correct—electrons are diffracted by crystal lattices just as if they are a form of wave. The discovery was made independently in 1927 by two groups, Davisson and a new collaborator, Lester Germer, in the U.S., and George Thomson (son of J.

*See Jammer, *op. cit.*

J.) and research student Alexander Reid, working in England and using a different technique. Through his failure to accept Elsasser's calculations at face value, Davisson missed his chance for solo glory and shared the 1937 Nobel Prize in Physics with Thomson for their independent studies in 1927. But that makes for a nice historical footnote, which even Davisson must have appreciated, and that neatly sums up the basic feature of quantum theory.

In 1906, J. J. Thomson had received the Nobel Prize for proving that electrons are particles; in 1937 he saw his son awarded the Nobel Prize for proving that electrons are waves. Both father and son were correct, and both awards were fully merited. Electrons are particles; electrons are waves. From 1928 onward, the experimental evidence for de Broglie's wave/particle duality became overwhelming. Other particles, including the proton and the neutron,* were subsequently found to possess wave properties, including diffraction, and in a series of beautiful experiments in the late 1970s and 1980s Tony Klein and colleagues at the University of Melbourne repeated some of the classic experiments that had established the wave theory of light in the nineteenth century, but using a beam of neutrons instead of a beam of light.†

A BREAK
WITH THE PAST

The complete break with classical physics comes with the realization that not just photons and electrons but all "particles" and all "waves" are in fact a mixture of wave and parti-

---

*Which was first detected only in 1932, by James Chadwick, who received a Nobel Prize as a result in 1935, a full two years before the similar recognition of the work of Davisson and Thomson.

†These experiments have potential for practical applications, including the possibility of a "neutron microscope." See *New Scientist*, 2 September, 1982, page 631.

cle. It just happens that in our everyday world the particle
component overwhelmingly dominates the mixture in the
case of, say, a bowling ball, or a house. The wave aspect is
still there, in accordance with the relation $p \lambda = h$, although
it is totally insignificant. In the world of the very small,
where particle and wave aspects of reality are equally significant, things do not behave in any way that we can understand from our experience of the everyday world. It isn't
just that Bohr's atom with its electron "orbits" is a false picture; *all* pictures are false, and there is no physical analogy
we can make to understand what goes on inside atoms.
Atoms behave like atoms, nothing else.

Sir Arthur Eddington summed up the situation bril-
liantly in his book *The Nature of the Physical World,* published in 1929. "No familiar conceptions can be woven
around the electron," he said, and our best description of
the atom boils down to "something unknown is doing we
don't know what." He notes that this "does not sound a
particularly illuminating theory. I have read something like
it elsewhere—

> The slithy toves
> Did gyre and gimbal in the wabe"

But the point is that although we do not know *what* electrons are doing in atoms, we do know that the number of
electrons is important. Adding a few numbers makes "Jabberwocky" scientific— "Eight slithy toves gyre and gimbal
in the oxygen wabe; seven in nitrogen . . . if one of its toves
escapes, oxygen will be masquerading in a garb properly
belonging to nitrogen."

This is not a facetious remark. Provided the numbers
are unchanged, as Eddington pointed out more than fifty
years ago, all the fundamentals of physics could be translated into "Jabberwocky." There would be no loss of meaning, and conceivably a great benefit if we broke the
instinctive association in our minds of atoms with hard
spheres and electrons with tiny particles. The point is
clearly made by the confusion surrounding a property of
the electron which is called "spin," but is nothing like the
behavior of a child's spinning top, or the rotation of the
earth on its axis as it orbits around the sun.

One of the puzzles of atomic spectroscopy that the simple Bohr model of the atom failed to explain involves the splitting of spectral lines that "ought" to be single into closely spaced multiplets. Because each spectral line is associated with a transition from one energy state to another, the number of lines in the spectrum reveals how many energy states there are in the atom—how many "steps" there are on the quantum staircase, and how deep each tread is. From their studies of spectra, the physicists of the early 1920s came up with several possible explanations for the multiplet structure. What proved to be the best explanation came from Wolfgang Pauli, and it involved assigning four separate quantum numbers to the electron. This was in 1924, when physicists still thought of the electron as a particle and tried to explain its quantum properties in terms familiar from the everyday world. Three of these numbers were already included in the Bohr model, and were thought of as describing the angular momentum of an electron (the speed with which it moved round its orbit), the shape of the orbit, and its orientation. The fourth number had to be associated with some other property of the electron, a property that came in only two varieties, to account for the observed splitting of the spectral lines.

It didn't take long for people to latch on to the idea that Pauli's fourth quantum number described the electron's "spin," which could be thought of as pointing either up or down, giving a nice double-valued quantum number. The first person to propose this was Ralph Kronig, a young physicist who was visiting Europe having just finished his PhD studies at Columbia University.* He proposed that the electron had an instrinsic spin of one half in the natural units $(h/2\pi)$, and that this spin could line up either parallel to the magnetic field of the atom or antiparallel.† To his surprise, Pauli himself strongly opposed the idea, largely because it could not be reconciled with the idea of the elec-

*Arthur Compton had, in fact, speculated that the electron might spin back in 1920, but this idea had been aired in a different context and Kronig was not aware of it.
†The 2π comes in because there are that many radians in a complete circle, 360°. The fundamental unit $h/2\pi$ is usually written as $\hbar$. More of this later.

tron as a particle within the framework of relativistic theory. Just as an electron in orbit around the nucleus "ought" not to be stable according to classical electromagnetic theory, so a spinning electron "ought" not to be stable according to relativity theory. Pauli should, perhaps, have been more open minded, but the result was that Kronig gave up the idea and never published it. Less than a year later, however, the same idea occurred to George Uhlenbeck and Samuel Goudsmit, of the Institute for Theoretical Physics, in Leyden. They published the suggestion in the German journal *Die Naturwissenschaften* late in 1925, and in *Nature* early in 1926.

The theory of the spinning electron was soon refined to explain fully the troublesome splitting of spectral lines, and by March 1926 Pauli himself was convinced. But what is this thing called spin? If you try to explain it in ordinary language, the concept, like so many quantum concepts, slips away. In one "explanation," for example, you might be told (correctly as far as it goes) that electron spin is not like the spin of a child's top because the electron has to spin around *twice* to get back to where it started. Then again, how can an electron wave "spin" at all? Nobody was happier than Pauli when Bohr was able to establish, in 1932, that electron spin cannot be measured by any classical experiment, such as deflection of beams of electrons by magnetic fields. It is a property that *only* appears in quantum interactions, such as the ones that produce the splitting of the spectral lines, and it has no classical meaning whatsoever. How much easier it might have been for Pauli and his colleagues struggling to understand the atom in the 1920s, if they had talked about the electron's "gyre" instead of its "spin" in the first place!

Alas, we are stuck with the term *spin* now, and no campaign for the abolition of classical terminology in quantum physics is likely to succeed. From here on, though, if ever you are pulled up short by the appearance of a familiar word in an unfamiliar context, try changing it into jabberwocky and see if that looks less frightening. *Nobody* understands what "really" goes on in atoms, but Pauli's four quantum numbers do explain some very crucial features of the way "slithy toves" fit into the different "wabes."

PAULI AND EXCLUSION

Wolfgang Pauli was one of the most remarkable of the remarkable body of scientists who founded quantum theory. Born in Vienna in 1900, he enrolled at the University of Munich in 1918, but brought with him a reputation as a precocious mathematician and a finished paper on general relativity theory that immediately aroused Einstein's interest, and was published in January 1919. Swallowing up physics from classes at the university and the Institute of Theoretical Physics, as well as his own reading, his command of relativity was so great that in 1920 he was assigned the task of writing a major review article on the subject for a definitive encyclopedia of mathematics. This masterly article by the twenty-year-old student spread his fame throughout the scientific community, where the work was praised highly by the likes of Max Born, whom he joined in Göttingen, as assistant, in 1921. From Göttingen he soon moved on, first to Hamburg and later to Bohr's Institute in Denmark. But Born did not suffer from the loss—his new assistant, Werner Heisenberg, was just as gifted and played a key role in the development of quantum theory.*

Even before Pauli's fourth quantum number was labeled "spin," he had been able, in 1925, to use the fact of the four numbers to resolve one of the great puzzles of the Bohr atom. In the case of hydrogen, the single electron naturally sits in the lowest energy state available, at the bottom

---

*See, for example, *The Born-Einstein Letters*. In a letter dated 12 February 1921, Born says, "Pauli's article for the Encyclopedia is apparently finished, and the weight of the paper is said to be 2½ kilos. This should give some indication of its intellectual weight. The little chap is not only clever but industrious as well." The clever little chap received his PhD in 1921, shortly before his brief spell as Born's assistant.

of the quantum staircase. If it is excited—by a collision, perhaps—it may jump up to a higher step on the staircase, then fall back to the ground state, emitting a quantum of radiation as it does so. But when more electrons are added to the system, for more massive atoms, they do not all fall into the ground state, but distribute themselves up the steps of the staircase. Bohr talked of the electrons as being in "shells" around the nucleus, with "new" electrons going into the shell with the least energy until it was full, and then into the next shell, and so on. In this way he built up the periodic table of the elements and explained many chemical mysteries. But he did not explain how or why a shell became full—why the first shell could contain only two electrons, but the next eight, and so on.

Each of Bohr's shells corresponded to a set of quantum numbers, and Pauli realized in 1925 that with the addition of his fourth quantum number for the electron the number of electrons in each full shell exactly corresponded to the number of *different* sets of quantum numbers belonging to that shell. He formulated what is now known as the Pauli Exclusion Principle, that no two electrons can have the same set of quantum numbers, and thereby provided a reason for the way the shells fill up for more and more massive atoms.

The exclusion principle, and the discovery of electron spin, really arrived ahead of their time, and were only fully fitted into the new physics in the late 1920s—after the new physics had itself been invented. Because of the almost headlong progress in physics in 1925 and 1926, the importance of exclusion sometimes gets overlooked, but it is, in fact, a concept as fundamental and far reaching as the concept of relativity, and it has broad applications across physics. The Pauli Exclusion Principle, it turns out, applies to *all* particles that have a half-integral amount of spin— $(1/2)\hbar$, $(3/2)\hbar$, $(5/2)\hbar$ and so on. Particles that have no spin at all (like photons) or integer spin ($\hbar$, $2\hbar$, $3\hbar$, and so on) behave in a completely different fashion, following a different set of rules. The rules that are obeyed by the half-spin particles are called Fermi-Dirac statistics, after Enrico Fermi and Paul Dirac, who worked them out in 1925 and

1926. Such particles are called "fermions." The rules obeyed by full-spin particles are called Bose-Einstein statistics, after the two men who worked them out, and the particles are called "bosons."

Bose-Einstein statistics were being developed at the same time, 1924–1925, as all the excitement about de Broglie waves, Compton effect, and electron spin. They mark Einstein's last great contribution to quantum theory (indeed, his last great piece of scientific work), and they too represent a complete break with classical ideas.

Satyendra Bose was born in Calcutta in 1894, and in 1924 he was Reader in Physics at the then new Dacca University. Following the work of Planck, Einstein, Bohr, and Sommerfeld from afar, and aware of the still imperfect basis of Planck's law, he set about deriving the blackbody law in a new way, starting out from the assumption that light comes in the form of photons, as they are now called. He came up with a very simple derivation of the law involving massless particles that obey a special kind of statistics, and sent a copy of his work, in English, to Einstein with a request that he should pass it on for publication in the *Zeitschrift für Physik*. Einstein was so impressed by the work that he translated it into German himself and passed it on with a strong recommendation, ensuring its publication in August 1924. By removing all elements of classical theory and deriving Planck's law from a combination of light quanta—regarded as relativistic particles with zero mass—and statistical methods, Bose finally cut quantum theory free from its classical antecedents. Radiation could now be treated as a quantum gas, and the statistics involved counting particles, not counting wave frequencies.

Einstein developed the statistics further, and applied them to what was then the hypothetical case of a collection of atoms—gas or liquid—obeying the same rules. The statistics turned out to be inappropriate for real gases at room temperature, but they are exactly right to account for the bizarre properties of superfluid helium, a liquid cooled close to the absolute zero of temperature, -273 °C. With Fermi-Dirac statistics coming on the scene by 1926, it took some time for physicists to sort out which rules applied where, and to appreciate the significance of the half-integer spin.

The subtleties need not concern us now, but the distinction between fermions and bosons is an important one that can be easily understood. Some years ago, I went to see a play starring the comedian Spike Milligan. Just before the curtain went up, the great man himself appeared on stage and took a baleful look at the handful of empty seats in the most expensive part of the auditorium, near the stage. "They'll never find anyone to buy these now," he said, "you might as well all move up here where I can see you." The audience did as he suggested—everybody moved forward so that all the seats near the stage were full, while the handful of empty seats was left at the back. We were acting like nice, well-behaved fermions, each person occupying just one seat (one quantum state) and filling up the seats from the most desirable "ground state," by the stage, outward.

Contrast this with the audience at a recent Bruce Springsteen concert I attended. There, every seat was full, but there was a small gap between the front row of seats and the stage. As the stage lights went up and the band hit the first chord of "Born to Run" the entire audience surged forward out of their seats and crammed up against the stage. All of the "particles" crammed into the same "energy state" indistinguishably—and that is the difference between fermions and bosons. Fermions obey the exclusion principle, bosons do not.

All the "material" particles that we are used to—electrons, protons, and neutrons—are fermions, and without the exclusion principle the variety of the chemical elements and all the features that make up our physical world would not exist. The bosons are more ghostly particles, such as photons, and the blackbody law is a direct result of all the photons trying to get into the same energy state. Helium atoms can mimic the properties of bosons, under the right conditions, and become superfluid because each atom of ^{4}He contains two protons and two neutrons, with their half-integer spins arranged to add up to zero. Fermions are also conserved in interactions between particles—it is impossi-

ble to increase the overall number of electrons in the universe—whereas bosons, as anyone who has ever switched a
light on knows, can be manufactured in vast quantities.

WHERE NEXT?

Although it all sounds reasonably neat and tidy from the
perspective of the 1980s, by 1925 quantum theory was in a
mess. There was no great highway of progress, but rather
many individuals each hacking a separate path through the
jungle. The top researchers knew this only too well, and
expressed their concern publicly; but the great leap forward was to come, with one exception, from the new generation who entered research after the First World War and
were, perhaps as a result, open to new ideas. In 1924, Max
Born commented that "at the moment one possesses only a
few unclear hints" about the way classical laws need to be
modified to explain atomic properties, and in his textbook
on atomic theory published in 1925 he promised a second
volume to complete the job, one that he thought would "remain unwritten for several years yet."*

Heisenberg, after an unsuccessful attempt to calculate
the structure of the helium atom, commented to Pauli early
in 1923, "What a misery!"—a phrase Pauli repeated in a
letter to Sommerfeld in July of that year, saying "The theory . . . with atoms having more than one electron, it is
such a great misery." In May 1925 Pauli wrote to Kronig
saying that "physics at the moment is again very muddled,"
and by 1925 Bohr himself was similarly gloomy about the
many problems that beset his model of the atom. As late as
June 1926, Wilhelm Wien, whose blackbody law had been
one of the springboards for Planck's leap in the dark, wrote
to Schrödinger about the "morass of integral and half-
integral quantum discontinuities and the arbitrary use of
the classical theory." All of the big names in quantum the-

*Quotes in this section taken from the epilogue to volume 1 of Mehra and
Rechenberg.

ory were aware of the problems—and all but one of the big names in quantum theory were alive in 1925 (the exception was Henri Poincaré; Lorentz, Planck, J. J. Thomson, Bohr, Einstein, and Born were still going strong, while Pauli, Heisenberg, Dirac, and others were beginning to make their mark). The two great authorities were Einstein and Bohr, but by 1925 they had begun to differ markedly in their scientific views. First, Bohr was one of the strongest opponents of the light quantum; then, as Einstein began to be concerned about the role of probability in quantum theory Bohr became its great champion. The statistical methods (ironically, introduced by Einstein) became the cornerstone of quantum theory, but as early as 1920 Einstein wrote to Born, "That business about causality causes me a lot of trouble, too . . . I must admit that . . . I lack the courage of my convictions," and the dialogue between Einstein and Bohr on this theme continued for thirty-five years, until Einstein's death.*

Max Jammer describes the situation at the beginning of 1925 as "a lamentable hodgepodge of hypotheses, principles, theorems and computational recipes."† Every problem in quantum physics had to be first "solved" using classical physics, and then be reworked by the judicious insertion of quantum numbers more by inspired guesswork than cool reasoning. The quantum theory was neither autonomous nor logically consistent, but existed as a parasite on classical physics, an exotic bloom without roots. No wonder Born thought that it would be years before he could write his second, definitive volume on atomic physics. And yet, it seems entirely in keeping with the strange story of the quantum that within a few months of the confused days of early 1925 the astonished scientific community was presented with not one but two complete, autonomous, logical and well-rooted quantum theories.

---

*Einstein expressed these doubts also in his correspondence with Born, published as *The Born-Einstein Letters*. The quote here is from page 23 of the Macmillan edition.

†*The Conceptual Development of Quantum Mechanics*, page 196.

CHAPTER SIX

MATRICES AND WAVES

Werner Heisenberg was born in Würzburg on 5 December 1901. In 1920 he entered the University of Munich, where he studied physics under Arnold Sommerfeld, one of the leading physicists of the time who had been closely involved with the development of the Bohr model of the atom. Heisenberg was plunged straight into research on quantum theory, and set the task of finding quantum numbers that could explain some of the splitting of spectral lines into pairs, or doublets. He found the answer in a couple of weeks—the whole pattern could be explained in terms of half-integer quantum numbers. The young, unprejudiced student had found the simplest solution to the problem, but his colleagues and his supervisor Sommerfeld were horrified. To Sommerfeld, steeped in the Bohr model, integral quantum numbers were established doctrine, and the young student's speculations were quickly quashed. The fear among the experts was that by introducing half integers into the equations they would open the door to quarter integers, then eighths and sixteenths, destroying

the fundamental basis of quantum theory. But they were
wrong.

Within a few months, the older and more senior physicist Alfred Landé came up with the same idea and published it; it later turned out that half-integer quantum
numbers are crucially important in the full quantum theory, and play a key role in describing the property of electrons called spin. Objects that have integer or zero spin,
like photons, obey the Bose-Einstein statistics, while those
that have half-integer spin (1/2, or 3/2, and so on) obey the
Fermi-Dirac statistics. The half-integer spin of the electron
is directly related to the structure of the atom and the periodic table of the elements. It is still true that quantum
numbers *change* only by whole integers, but a jump from
1/2 to 3/2, or 5/2 to 9/2, is just as legitimate as a jump from
1 to 2, or 7 to 12. So Heisenberg missed a chance for credit
for a new idea in quantum theory; but the point of the story
is that just as it took young men in the previous generation
to develop the first quantum theory, so in the 1920s it was
time again for young minds unencumbered by ideas that
"everyone knows" must be right to take the next step forward. Heisenberg certainly made up for missing out on one
minor scientific "first" with his work over the next few
years.

After a term working in Göttingen under Born, where
he had attended the famous "Bohr Festival," Heisenberg
returned to Munich and completed his PhD in 1923—still
not quite twenty-two years old. At that time, Wolfgang
Pauli, a close friend of Heisenberg's, equally precocious and
another former student of Sommerfeld's, was just moving
on from a spell as Born's assistant in Göttingen, and
Heisenberg took over the post in 1924. It was a job that
gave him the opportunity to work for several months with
Bohr in Copenhagen, and by 1925 the precocious mathematical physicist was better equipped than anyone to find
the logical quantum theory that every physicist expected to
be found eventually, but no one expected to find so soon.

Heisenberg's breakthrough was founded on an idea he
picked up from the Göttingen group—nobody now is quite
sure who suggested it first—that a physical theory should

only be concerned with things that can actually be observed by experiments. This sounds trite, but it is actually a
very deep insight. An experiment that "observes" electrons
in atoms, for example, doesn't show us a picture of little
hard balls orbiting around the nucleus—there is no way to
observe the *orbit,* and the evidence from spectral lines tells
us what happens to electrons when they move from one
energy state (or orbit, in Bohr's language) to another. All of
the observable features of electrons and atoms deal with
two states, and the concept of an orbit is something tacked
on to the observations by analogy with the way things move
in our everyday world (remember the slithy toves). Heisenberg stripped away the clutter of the everyday analogies,
and worked intensively on the mathematics that described
not one "state" of an atom or electron, but the associations
between *pairs* of states.

BREAKTHROUGH IN HELIGOLAND

The story is often told of how Heisenberg was struck down
by a severe bout of hayfever in May 1925, and went off to
recuperate on the rocky island of Heligoland, where he
painstakingly tackled the task of interpreting what was
known about quantum behavior in these terms. With no
distractions on the island, and his hayfever gone, Heisenberg was able to work intensively on the problem. In his
autobiographical *Physics and Beyond,* he described his
feelings as the numbers began to fall into place, and how at
three o'clock one morning he "could no longer doubt the
mathematical consistency and coherence of the kind of
quantum mechanics to which my calculations pointed. At
first, I was deeply alarmed. I had the feeling that, through
the surface of atomic phenomena, I was looking at a
strangely beautiful interior, and felt almost giddy at the
thought that I now had to probe this wealth of mathematical structures nature had so generously spread out before me."

Returning to Göttingen, Heisenberg spent three weeks preparing his work in a form suitable for publication and sent a copy of the paper first to his old friend Pauli, asking if he thought it made sense. Pauli was enthusiastic, but Heisenberg was exhausted by his efforts and not yet sure that the work was ready for publication. He left the paper with Born to dispose of as he felt appropriate, and departed, in July 1925, to give a series of lectures in Leyden and Cambridge. Ironically, he did not choose to speak about his new work to the audiences there, who had to wait for news to reach them by other channels.

Born was happy enough to send Heisenberg's paper off to the *Zeitschrift für Physik,* and almost immediately realized what it was that Heisenberg had stumbled upon. The mathematics involving two states of an atom couldn't be dealt with by ordinary numbers, but involved arrays of numbers, which Heisenberg had thought of as tables. The best analogy is with a chess board. There are 64 squares on the board, and in this case you could identify each square by one number, in the range 1 to 64. However, chess players prefer to use a notation that labels the "columns" of squares across the board by the letters a, b, c, d, e, f, g, and h, with the "rows" numbered up the board 1, 2, 3, 4, 5, 6, 7, 8. Now, each square on the board can be identified by a unique pair of identifying labels: a1 is the home square of a rook; g2 is the home square of a knight's pawn, and so on. Heisenberg's tables, like a chess board, involved two-dimensional arrays of numbers, because he was doing calculations involving two states and their interactions. Those calculations involved, among other things, multiplying two such sets of numbers, or arrays, together, and Heisenberg had laboriously worked out the right mathematical tricks to do the job. But he had come up with a very curious result, so puzzling that it was one of the reasons for his diffidence about publishing his calculations. When two of these arrays are multiplied together, the "answer" you get depends on the order in which you do the multiplication.

This is strange indeed. It is as if 2×3 is not the same as 3×2, or in algebraic terms $a \times b \neq b \times a$. Born worried at this peculiarity day and night, convinced that something fundamental lay behind it. Suddenly, he saw the light. The mathematical arrays and tables of numbers, so laboriously constructed by Heisenberg, were *already known* in mathematics. A whole calculus of such numbers existed; they were called matrices, and Born had studied them in the early years of the twentieth century, when he was a student in Breslau. It isn't really surprising that he should have remembered this obscure branch of mathematics more than twenty years later, for there is one fundamental property of matrices that always makes a deep impression on students

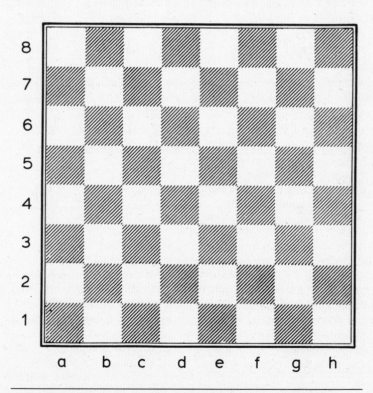

Figure 6.1/Each square on a chessboard can be identified by a paired number and letter, such as b4 or f7. Quantum mechanical states are also defined by pairs of numbers.

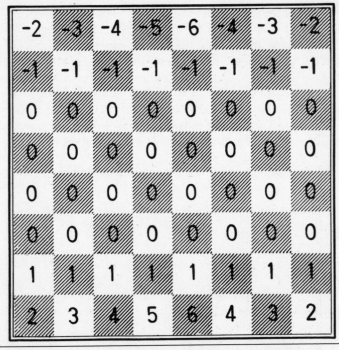

*Figure 6.2/*The "state" of each square on the chessboard is
determined by the chess piece occupying that square. In this
notation, a pawn is denoted by 1, rook by 2, and so on; positive
numbers are white pieces, negative black. We can describe a
change in the state of the whole board by an expression such
as "pawn to queen four," or by the algebraic notation e2–e4.
Quantum transitions are described in similar notation linking
paired (initial and final) states; in neither case do we have any
indication of how the transition from one state to another is
carried out, a point brought out most strongly by the knight's
move and by castling. In the chess analogy, we might fancifully
imagine the smallest possible change in the board, e2–e3,
as corresponding to the input of a quantum of energy,
$h\nu$, while the "transition" e3–e2 would then correspond to the
release of the same quantum of energy. The analogy is inexact
but highlights the way different forms of notation describe the
same event. Heisenberg, Dirac, and Schrödinger similarly found
different forms of mathematical notation to describe the same
quantum events.

when they first learn of it—the answer you get when you multiply matrices depends on the order in which you do the multiplying, or in mathematical language, matrices do not commute.

QUANTUM MATH

In the summer of 1925, working with Pascual Jordan, Born developed the beginnings of what is now known as matrix mechanics, and when Heisenberg returned to Copenhagen in September he joined them through correspondence in producing a comprehensive scientific paper on quantum mechanics. In this paper, far more clearly and explicitly than in Heisenberg's original paper, the three authors stressed the fundamental importance of the noncommutativity of quantum variables. Already, in his joint paper with Jordan, Born had found the relation $\mathbf{pq} - \mathbf{qp} = \hbar/i$, where $\mathbf{p}$ and $\mathbf{q}$ are matrices representing quantum variables, the equivalent in the quantum world of momentum and position. Planck's constant appears in the new equation, along with i, the square root of minus one; in what became known as the "three-man paper," the Göttingen team stressed that this is the "fundamental quantum-mechanical relation." But what did it mean in physical terms? Planck's constant was by now familiar enough, and physicists knew equations involving i (a clue to what was to come, if they had but realized it, since such equations generally involve oscillations, or waves). But matrices were totally unfamiliar to most mathematicians and physicists in 1925, and the noncommutativity seemed as strange to them as Planck's introduction of h had seemed at first sight to their predecessors in 1900. For those who could handle the mathematics, the results were dramatic. The equations of Newtonian mechanics were replaced by similar equations involving matrices and, says Heisenberg, "It was a strange experience to find that many of the old results of Newtonian mechanics, like conservation of energy, etc., could be derived also in the new scheme."[*] In other words, matrix mechanics *included* Newtonian mechanics within itself, just as Einstein's relativistic equations include Newton's equations as a special case. Unfortunately, few people could understand the mathematics, and it was not immediately appreciated by most physicists just how significant a breakthrough Heisenberg and the Göttingen group had made. There was one exception, however, and that was in Cambridge, England.

Paul Dirac was a few months younger than Heisenberg, having been born on 8 August 1902. He is generally regarded as the only English theorist who can rank with Newton, and he developed the most complete form of what is now called quantum mechanics. Yet he did not turn to theoretical physics until after he graduated from Bristol University in 1921, with a degree in engineering. Unable to find a position in engineering, he was offered a studentship to study mathematics in Cambridge, but was unable to take it up through lack of money. Staying in Bristol and living with his parents, he took the three-year course in mathematics in only two years, thanks to his engineering degree, and completed a BA in applied mathematics in 1923. Now, at last he could go to Cambridge to take up research, supported by a grant from the Department of Scientific and Industrial Research—and only on arriving in Cambridge did he learn for the first time about quantum theory.

So it was as an unknown and inexperienced research student that Dirac heard Heisenberg talk in Cambridge in July 1925. Although Heisenberg did not talk publicly about his new work then, he did mention it to Ralph Fowler, Dirac's supervisor, and as a result he sent Fowler a copy of the proof of the paper in the middle of August, before it appeared in the *Zeitschrift*. Fowler passed the paper on to Dirac, who had it in front of him before anybody outside Göttingen (except Heisenberg's friend Pauli) had had a chance to study the new theory. In this first paper, although he pointed out the noncommutativity of the variables in quantum mechanics—the matrices—Heisenberg did not

---

[*]*Physics and Philosophy,* page 41.

develop the idea, but tried to fudge around it. When he got
to grips with the equations, Dirac soon appreciated the *fundamental* importance of the simple fact that $a \times b \neq b \times a$.
Unlike Heisenberg, Dirac already knew of mathematical
quantities that behaved in this fashion, and within a few
weeks he was able to rework Heisenberg's equations in
terms of a branch of mathematics developed by William
Hamilton a century earlier. By one of the most delicious
scientific ironies, the Hamiltonian equations that proved so
useful in the new quantum theory, which dispensed altogether with electron orbits, had been developed in the
nineteenth century largely as an aid to the calculation of
the orbits of bodies in a system, like the solar system, where
there are several interacting planets.

So Dirac discovered, independently of the Göttingen
group, that the equations of quantum mechanics have the
same mathematical structure as the equations of classical
mechanics, and that classical mechanics is included within
quantum mechanics as a special case, corresponding to
large quantum numbers or to setting Planck's constant
equal to zero. Following his own direction, Dirac developed
yet another way of expressing the dynamics mathematically, using a special form of algebra, which he called
quantum algebra, involving the addition and multiplication
of quantum variables, or "q numbers." These q numbers
are strange beasts, not least because in this mathematical
world developed by Dirac it is impossible to say which of
two numbers a and b is bigger—the concept of one number
being bigger or smaller than another has no place in this
algebra. But, again, the rules of that mathematical system
exactly fitted the observations of the behavior of atomic processes. Indeed, it is correct to say that quantum algebra
includes matrix mechanics within itself, but does much
more besides.

Fowler immediately appreciated the importance of Dirac's work, and at his instigation it was published in the
Proceedings of the Royal Society in December 1925.
Among other things, the paper included, as an essential
component of the new theory, the half-integral quantum
numbers that had troubled Heisenberg a few years before.

Heisenberg, sent a copy of the manuscript of the paper by Dirac, was generous in his praise: "I have read your extraordinarily beautiful paper on quantum mechanics with the greatest interest, and there can be no doubt that all your results are correct . . . [the paper is] really better written and more concentrated than our attempts here."* In the first half of 1926, Dirac carried this work through in a series of four definitive papers, with the whole package forming the thesis for which he was duly awarded his doctorate. While all this was going on, Pauli had used matrix methods to predict, correctly, the Balmer series for the hydrogen atom, and by the end of 1925 it had become clear that the splitting of some spectral lines into doublets could indeed be best explained by assigning the new property called spin to the electron. The pieces fit together very well indeed, and the different mathematical tools used by the different exponents of matrix mechanics were clearly just different aspects of the same reality.†

Again, the game of chess can help to make this clear. There are several different ways of describing a chess game on the printed page. One way is to print a representative "chessboard" with the positions of all the pieces marked— but that would take up a lot of space if we wanted to record a whole game. Another way is to name the pieces being moved: "King's pawn to King's pawn four." And in the most concise algebraic notation the same move becomes simply "d2 − d4." Three different descriptions provide the same information about a real event, the transition of a pawn from one "state" to another (and, just as in the quantum world, we know *nothing* about how the pawn got from one state to the other, a point that is even more clear if you consider the knight's move). The different formulations of quantum mechanics are like this. Dirac's quantum algebra is the most

---

*Quoted by Mehra and Rechenberg, volume 4, page 159.
†In Dirac's version of quantum mechanics, a key expression in the Hamiltonian equations is replaced by the quantum mechanical expression $(ab - ba)/i\hbar$, which is just another form of the expression Born, Heisenberg, and Jordan called the "fundamental quantum-mechanical relation," in their three-man paper, written before Dirac's first paper on quantum mechanics appeared, but published after Dirac's paper.

elegant and "beautiful" in the mathematical sense; the matrix methods developed by Born and his collaborators in the wake of Heisenberg are more clumsy but none the less effective.*

Some of Dirac's most dramatic early results came when he tried to include special relativity in his quantum mechanics. Quite happy with the idea of light as a particle (the photon), Dirac was delighted to find that by including time as a q number along with all the rest in his equations he was inevitably led to the "prediction" that an atom must suffer a recoil when it emits light, just as it should do if the light is in the form of a particle carrying its own momentum, and he went on to develop a quantum-mechanical interpretation of the Compton effect. Dirac's calculations went in two parts, first the numerical manipulations involving the q numbers, and secondly the interpretation of the equations in terms of what might be physically observed. This process exactly fits the way nature seems to "make the calculation" and then present us with an observed event—an electron transition, say—but unfortunately instead of following this idea through completely in the years after 1926 physicists were seduced away from quantum algebra by the discovery of yet another mathematical technique that could solve the long-standing problems of quantum theory—wave mechanics. Matrix mechanics, and quantum algebra, had started out from the picture of an electron as a particle making a transition from one quantum state to another. But what about de Broglie's suggestion that electrons, and other particles, must also be thought of as waves?

---

*With characteristic, and genuine, modesty Dirac has described how easy it was to make progress once it was known that the correct quantum equations were simply classical equations put into Hamiltonian form. For any of the little puzzles that beset quantum theory, all that was necessary was to set up the equivalent classical equations, turn them into Hamiltonians, and solve the puzzle. "It was a game, a very interesting game one could play. Whenever one solved one of the little problems, one could write a paper about it. It was very easy in those days for any second-rate physicist to do first-rate work. There has not been such a glorious time since then. It is very difficult now for a first-rate physicist to do second-rate work." (*Directions in Physics*, page 7.)

SCHRÖDINGER'S THEORY

While matrix mechanics and quantum algebra were making their relatively unsung debuts on the scientific scene, there was plenty of other activity in the field of quantum theory. It seems European science had been bubbling with a ferment of ideas whose time had come, and different ideas popped up in different places, not necessarily in what now seems a logical order, and many of them were "discovered" by different people at around the same time. By late 1925, de Broglie's theory of electron waves had already appeared on the scene, but the definitive experiments that proved the wave nature of the electron had not yet been carried out. Quite independently of the work of Heisenberg and his colleagues, this led to another discovery, a quantum mathematics based on the wave idea.

The idea came from de Broglie, via Einstein. De Broglie's work might have remained obscure for years, regarded as no more than an interesting mathematical quirk with no physical reality, if it hadn't come to Einstein's attention. It was Einstein who told Born about the idea and thereby set off the train of experimental work that proved the reality of electron waves; and it was in one of Einstein's papers, published in February 1925, that Erwin Schrödinger read Einstein's comment on de Broglie's work, "I believe that it involves more than merely an analogy." In those days, physicists hung on Einstein's every word, and a nod from the great man was enough to set Schrödinger off on an investigation of the implications of taking de Broglie's idea at face value.

Schrödinger is the odd one out among the physicists who developed the new quantum theory. He was born in 1887 and was thirty-nine years old when he completed his greatest contribution to science—a remarkably advanced age for original scientific work of such importance. He had received his doctorate back in 1910, and since 1921 he had been Professor of Physics in Zürich, a pillar of scientific respectability, and not an obvious source of revolutionary new ideas. But, as we shall see, the nature of his contribution to quantum theory was very much what we might expect from a member of the older generation in the mid-1920s. Where the Göttingen group, and Dirac even more so, made quantum theory more abstract and cut it free from everyday physical ideas, Schrödinger tried to restore easily understood physical concepts, describing quantum physics in terms of waves, which are familiar features of the physical world, and fighting to the end of his life against the new ideas of indeterminacy and the instantaneous jumping of electrons from one state to another. He gave physics an invaluable practical tool for solving problems, but in conceptual terms his wave mechanics was a step backward, a return to nineteenth-century ideas.

De Broglie had pointed the way with his idea that electron waves "in orbit" around an atomic nucleus had to fit a whole number of wavelengths into each orbit, so that in-between orbits were "forbidden." Schrödinger used the mathematics of waves to calculate the energy levels allowed in such a situation, and was disappointed at first to get answers that did not agree with the known patterns of atomic spectra. In fact, there was nothing wrong with his technique, and the only reason for his initial failure was that he had not taken account of the spin of the electron—hardly surprising, since at that time in 1925 the concept of electron spin had not yet emerged. So he put the work aside for several months, and thereby missed being the first person to publish a complete, logical, and consistent mathematical treatment of quanta. He came back to the idea when he was asked to give a colloquium explaining de Broglie's work, and it was then that he found that if he left out the relativistic effects from his calculations he could get a good agreement with observations of atoms in situations where relativistic effects were not important. As Dirac was later to show, electron spin is essentially a relativistic property (and nothing like the property called spin associated with rotating objects in the everyday world). So Schrödinger's great contribution to quantum theory was published in a series of papers in 1926, hot on the heels of the papers from Heisenberg, Born and Jordan, and Dirac.

The equations in Schrödinger's variation on the quantum theme are members of the same family of equations that describe real waves in the everyday world—waves on the surface of the ocean, or the sound waves that carry noises through the atmosphere. The world of physics greeted them with enthusiasm, precisely because they seemed so comfortable and familiar. No two approaches to the problem could have been more different. Heisenberg deliberately discarded any picture of the atom and dealt only in terms of quantities that could be measured by experiment; at the heart of his theory, though, was the idea that electrons are particles. Schrödinger started out from a clear physical picture of the atom as a "real" entity; at the heart of his theory was the idea that electrons are waves. Both approaches produced sets of equations that exactly described the behavior of things that could be measured in the quantum world.

At first sight, this was astonishing. Yet before long Schrödinger himself, the American Carl Eckart, and then Dirac proved mathematically that the different sets of equations were in fact exactly equivalent to one another, different views of the same mathematical world. Schrödinger's equations include both the noncommutativity relation, and the crucial factor $\hbar/i$, in essentially the same way that they turn up in matrix mechanics and quantum algebra. The discovery that the different approaches to the problem were indeed mathematically equivalent to one another strengthened the confidence of physicists in them all. It seems that, whatever sort of mathematical formalism you like to use, when you tackle the fundamental problems of quantum theory you are led inexorably to the same "answers." Mathematically speaking, Dirac's variation on the theme is the most complete, since his quantum algebra includes both matrix mechanics and wave mechanics as special cases. Naturally enough, however, the physicists of the 1920s chose to use the most familiar version of the equations, Schrödinger's waves, which they could understand in everyday terms, and whose equations were familiar from the problems of everyday physics—optics, hydrodynamics, and the like. But the very success of Schrödinger's version of the story may have held back any fundamental understanding of the quantum world for decades.

A BACKWARD STEP

With hindsight, it seems surprising that Dirac did not discover (or invent) wave mechanics, for the very equations developed by Hamilton that proved so useful in quantum mechanics had their origin in a nineteenth-century attempt to unify the wave and particle theories of light. Sir William Hamilton was born in Dublin in 1805, and became regarded by many as the foremost mathematician of his age. His greatest achievement (although not regarded as such at the time) was a unification of the laws of optics and dynamics in one mathematical framework, one set of equations that could be used to describe both the motion of a wave and the motion of a particle. The work was published in the late 1820s and early 1830s, and both aspects were taken up by others. The mechanics and the optics were each useful to researchers in the second half of the nineteenth century, but scarcely anybody took notice of the coupled mechanics/optics system that was Hamilton's real concern. The clear implication of Hamilton's work is that just as light "rays" have to be replaced by the concept of waves in optics, so particle tracks have to be replaced by wave motions in mechanics. But such an idea would have been so alien to nineteenth-century physics that nobody—not even Hamilton—articulated it. It wasn't that the idea was raised and rejected as absurd; it was literally too bizarre even to occur to anyone. This would have been an impossible conclusion for any nineteenth-century physicist to have reached, and it was inevitable that the idea would only become established after the inadequacy of classical mechanics as a description of atomic processes had been proved. But, bearing in mind that he also invented the form of mathematics in which $a \times b \neq b \times a$, it would be no exaggeration to describe Sir William Hamilton as the forgotten founder of quantum mechanics. Had he been around at the time, he would have been quick to see the connection between matrix mechanics and wave mechanics; Dirac could have done so, but it isn't really surprising that he missed the connection at first. He was, after all, a student deep in his first major piece of research, and there is a limit to how much one man can do in a few weeks. Perhaps more importantly, though, he was dealing with abstract ideas, and following up Heisenberg's attempt to cut quantum physics free from the cozy everyday picture of electrons orbiting atomic nuclei, and had no expectation of finding a nice, intuitive physical picture of the atom. What people did not immediately appreciate was that wave mechanics itself did not, in spite of Schrödinger's expectations, provide such a cozy picture.

Schrödinger thought that he had eliminated quantum jumps from one state to another by putting waves into quantum theory. He envisaged the "transitions" of an electron from one energy state to another as something like the change in the vibration of a violin string from one note to another (one harmonic to another), and he thought of the wave in his wave equation as the matter wave invoked by de Broglie. But as other researchers sought to find the underlying significance of the equations, these hopes of restoring classical physics to the center stage evaporated. Bohr, for example, was baffled by the wave concept. How could a wave, or a set of interacting waves, make a Geiger counter click just as if it recorded a single particle? What was actually "waving" in the atom? And, crucially, how could the nature of blackbody radiation be explained in terms of Schrödinger's waves? So in 1926 Bohr invited Schrödinger to spend some time in Copenhagen, where they tackled these problems and came up with solutions that were not very tasteful to Schrödinger.

First, the waves themselves turned out, on close inspection, to be as abstract as Dirac's q numbers. The mathematics showed that they couldn't be real waves in space, like ripples on a pond, but represented a complex form of vibrations in an imaginary mathematical space called configuration space. Worse than that, each particle (each electron, say) needs its own three dimensions. One electron on its own can be described by a wave equation in three-dimensional configuration space; to describe two electrons requires a six-dimensional configuration space; three electrons require nine dimensions, and so on. As for the blackbody radiation, even when everything was converted into wave-mechanical language the need for discrete quanta, and quantum jumps, remained. Schrödinger was disgusted, and made the remark which has often been quoted, with slight variations in the translation: "Had I known that we were not going to get rid of this damned quantum jumping, I never would have involved myself in this business." As Heisenberg put it in his book *Physics and Philosophy*, ". . . The paradoxes of the dualism between wave picture and particle picture were not solved; they were hidden somehow in the mathematical scheme."

Without doubt, the appealing picture of physically real waves circling around atomic nuclei that had led Schrödinger to discover the wave equation that now bears his name is wrong. Wave mechanics is no more a guide to the reality of the atomic world than matrix mechanics, but unlike matrix mechanics, wave mechanics gives an *illusion* of something familiar and comfortable. It is that cozy illusion that has persisted to the present day and that has disguised the fact that the atomic world is totally different from the everyday world. Several generations of students, who have now grown up to be professors themselves, might have achieved a much deeper understanding of quantum theory if they had been forced to come to grips with the abstract nature of Dirac's approach, rather than being able to imagine that what they knew about the behavior of waves in the everyday world gave a picture of the way atoms behave. And that is why it seems to me that although there have been enormous strides in the application of quantum mechanics, cookbook fashion, to many interesting problems (remember Dirac's remark about the second-rate physicists doing first-rate work), we are scarcely today, more than fifty years on, any better placed than the physicists of the late 1920s concerning our fundamental understanding of quantum physics. The very success of the

Schrödinger equation as a practical tool has stopped people from thinking deeply about how and why the tool works.

QUANTUM COOKERY

The basics of quantum cookery—practical quantum physics since the 1920s—depend upon ideas developed by Bohr and Born in the late 1920s. Bohr gave us a philosophical basis with which to reconcile the dual particle/wave nature of the quantum world, and Born gave us the basic rules to follow in preparing our quantum recipes.

Bohr said that *both* the theoretical pictures, particle physics and wave physics, are equally valid, complementary descriptions of the same reality. Neither description is complete in itself, but there are circumstances where it is more appropriate to use the particle concept, and circumstances where it is better to use the wave concept. A fundamental entity such as an electron is neither a particle nor a wave, but under some circumstances it behaves as if it were a wave, and under other circumstances it behaves as if it were a particle (really, of course, it is a slithy tove). But under no circumstances can you invent an experiment that will show the electron behaving in both fashions at once. This idea of wave and particle being two complementary facets of the electron's complex personality is called complementarity.

Born found a new way of interpreting Schrödinger's waves. The important thing in Schrödinger's equation that corresponds to the physical ripples on the pond in the everyday world is a wave function, which is usually denoted by the Greek letter psi (ψ). Working in Göttingen alongside experimental physicists who were performing new electron experiments confirming the particle nature of the electron almost every day, Born simply could not accept that this psi function corresponded to a "real" electron wave, although like almost all physicists at the time (and since) he found the wave equations the most convenient for solving many

problems. So he tried to find a way of associating a wave
function with the existence of particles. The idea he picked
up on was one that had been aired before in the debate
about the nature of light, but which he now took over and
refined. The particles were real, said Born, but in some
sense they were guided by the wave, and the strength of the
wave (more precisely, the value of ψ^2) at any point in space
was a measure of the *probability* of finding the particle at
that particular point. We can never know for sure where a
particle like an electron is, but the wave function enables
us to work out the probability that, when we carry out an
experiment designed to locate the electron, we will find it in
a certain place. The strangest thing about this idea is that it
means that any electron might be anywhere at all, it's just
that it is extremely likely to be in some locations and very
unlikely to be in others. But like the statistical rules that say
that it is *possible* for all the air in the room to gather in the
corners, so Born's interpretation of ψ removed some of the
certainty from the already uncertain quantum world.

The ideas of both Bohr and Born tied in very well with
Heisenberg's discovery, late in 1926, that uncertainty is indeed inherent in the equations of quantum mechanics. The
mathematics that says that $\mathbf{pq} \neq \mathbf{qp}$ also says that we can
never be certain just what $\mathbf{p}$ and $\mathbf{q}$ are. If we call $\mathbf{p}$ the
momentum of, say, an electron, and use $\mathbf{q}$ as a label of its
position, we can imagine measuring either $\mathbf{p}$ or $\mathbf{q}$ very accurately. The amount of "mistake" in our measurement might
be called $\Delta\mathbf{p}$ or $\Delta\mathbf{q}$, since mathematicians use the Greek
letter delta, Δ, to symbolize small pieces of variable quantities. What Heisenberg showed was that if you tried, in this
case, to measure *both* the position and momentum of an
electron you could never quite succeed, because $\Delta\mathbf{p} \times \Delta\mathbf{q}$
must *always* be bigger than $\hbar$, Planck's constant divided by
2π. The more accurately we know the position of an object,
the less certain we are of its momentum—where it is going.
And if we know its momentum very accurately, then we
can't be quite sure where it is. This uncertainty relation has
far-reaching implications that are discussed in Part Three
of this book. The important point to appreciate, however, is
that it does not represent any deficiency in the experiments

used to measure the properties of the electron. It is a car-
dinal rule of quantum mechanics that *in principle* it is impossible to measure precisely certain pairs of properties,
including position/momentum, simultaneously. There is no
absolute truth at the quantum level.*

Heisenberg's uncertainty relation measures the
amount by which the complementary descriptions of the
electron, or other fundamental entities, overlap. Position is
very much a particle property—particles can be located
precisely. Waves, on the other hand, have no precise loca-
tion, but they do have momentum. The more you know
about the wave aspect of reality, the less you know about
the particle, and vice versa. Experiments designed to detect
particles always detect particles; experiments designed to
detect waves always detect waves. No experiment shows
the electron behaving like a wave and a particle at the same
time.

Bohr stressed the importance of experiments in our understanding of the quantum world. We can only probe the
quantum world by doing experiments, and each experiment, in effect, asks a question of the quantum world. The
questions we ask are highly colored by our everyday experience, so that we seek properties like "momentum" and
"wavelength," and we get "answers" that we interpret in
terms of those properties. The experiments are rooted in
classical physics, even though we know that classical physics does not work as a description of atomic processes. In
addition, we have to interfere with the atomic processes in
order to observe them at all, and, said Bohr, that means that
it is meaningless to ask what the atoms are doing when we

---

*In the everyday world, the same uncertainty relation applies, but because p
and q are so much bigger than $\hbar$ the amount of uncertainty involved is only a
tiny fraction of the equivalent macroscopic property. Planck's constant, h, is
about 6.6×10^{-27}, and π is a bit bigger than three. In round terms, therefore,
$\hbar$ is just about 10^{-27}. We can measure the position and the momentum of a
pool ball as accurately as we like by tracking it as it rolls across a table, and the
natural uncertainty of something comparable to 10^{-27} in either position or
momentum won't show up in any practical way. As always, the quantum
effects only become important if the numbers in the equations are about the
same size as, or smaller than, Planck's constant.

are not looking at them. All we can do, as Born explained, is to calculate the probability that a particular experiment will come up with a particular result.

This collection of ideas—uncertainty, complementarity, probability, and the disturbance of the system being observed by the observer—are together referred to as the "Copenhagen interpretation" of quantum mechanics, although nobody in Copenhagen (or anywhere else) ever set down in so many words a definitive statement labeled "the Copenhagen interpretation," and one of the key ingredients, the statistical interpretation of the wave function, actually came from Max Born in Göttingen. The Copenhagen interpretation is many things to many people, if not quite all things to all men, and itself has a slipperiness appropriate in the slippery world of quantum mechanical slithy toves. Bohr first presented the concept in public at a conference in Tomo, Italy, in September 1927. That marked the completion of the consistent theory of quantum mechanics in a form where it could be used by any competent physicist to solve problems involving atoms and molecules, with no great need for thought about the fundamentals but a simple willingness to follow the recipe book and turn out the answers.

In the following decades, many fundamental contributions were made by the likes of Dirac and Pauli, and the pioneers of the new quantum theory were duly honored by the Nobel Committee, although the allocation of the awards followed the committee's own curious logic. Heisenberg received his in 1932, and was mortified that the prize had not gone also to his colleagues Born and Jordan; Born himself remained bitter about this for years, often commenting that Heisenberg didn't even know what a matrix was until he (Born) had told him, and writing to Einstein in 1953 "in those days he actually had no idea what a matrix was. It was he who reaped all the rewards of our work together, such as the Nobel Prize."* Schrödinger and Dirac shared the physics prize in 1933, but Pauli had to wait until 1945

---

*Born-Einstein Letters, page 203.

to receive his, for the discovery of the exclusion principle, and Born was at last honored in 1954 with a Nobel Prize for his work on the probabilistic interpretation of quantum mechanics.*

Yet all of this activity—the new discoveries of the 1930s, the award of the prizes, and the new applications of quantum theory in the decades after World War II, should not disguise the fact that the era of fundamental advances was, for the time being, over. It may be that we are on the brink of another such era, and that new progress will be made by discarding the Copenhagen interpretation and the cozy pseudofamiliarity of Schrödinger's wave function. Before we look at these dramatic possibilities, though, it is only fair to spell out just how much has been achieved with the theory that was essentially complete before the end of the 1920s.

---

*Not before time, in his opinion (and, to be fair, that of many others). In the *Born-Einstein Letters*, he recalls (page 229) that "the fact that I did not receive the Nobel Prize in 1932 together with Heisenberg hurt me very much at the time, in spite of a kind letter from Heisenberg." He explains the delay in receiving acknowledgment of his work on the statistical interpretation of the wave equation as due to the opposition of Einstein, Schrödinger, Planck, and de Broglie, to the idea—certainly not names for the Nobel Committee to dismiss lightly—and he makes passing reference to the "Copenhagen school, which today lends its name almost everywhere to the line of thinking I originated," meaning the Copenhagen interpretation incorporating the statistical ideas. These aren't just the crusty remarks of an old man, but have substantial foundation; everybody in the quantum-mechanical trade was delighted by the belated recognition of Born's contribution. Nobody more so than Heisenberg, who later commented to Jagdish Mehra, "I was *so relieved* when Born was awarded the Nobel Prize." (Mehra and Rechenberg, volume 4, page 281.)

CHAPTER SEVEN

COOKING WITH QUANTA

In order to use the recipes in the quantum cookbook, physicists need to know a few simple things. There is no model of what the atom and elementary particles are really like, and nothing that tells us what goes on when we are not looking at them. But the equations of wave mechanics (the most popular and widely used variation on the theme) can be used to make predictions on a statistical basis. If we make an observation of a quantum system and get the answer A to our measurement, then the quantum equations tell us what the probability is of getting answer B (or C, or D, or whatever) if we make the same observation a certain time later. Quantum theory does *not* say what atoms are like, or what they are doing when we are not looking at them. Unfortunately, most of the people who use the wave equations today do not appreciate this, and only pay lip service to the role of probabilities. Students learn what Ted Bastin has called "a crystalized form of the play of ideas current in the late twenties . . . what the average physicist who never actually asks himself what he believes on foundational questions, is able to work with in solving his detailed problems."* They learn to think of the waves as real, and few of them get through a course in quantum theory without coming away with a picture of the atom in their imagination. People work with the probabilistic interpretation without really understanding it, and it is a testimony to the power of the equations developed by Schrödinger and Dirac in particular, and the interpretation provided by Born, that even without understanding why the recipes work people are able to cook so effectively with quanta.

The first quantum chef was Dirac. Just as he had been the first person outside Göttingen to understand the new matrix mechanics and develop it further, so he was the person who took Schrödinger's wave mechanics and put it on a more secure footing while developing it further. In adapting the equations to fit the requirements of relativity theory, adding time as the fourth dimension, Dirac found in 1928 that he had to introduce the term that is now regarded as representing the electron's spin, unexpectedly providing the explanation of the doublet splitting of spectral lines that had baffled theorists throughout the decade. The same improvement of the equations threw up another unexpected result, one that opened the way for the modern development of particle physics.

ANTIMATTER

According to Einstein's equations, the energy of a particle that has mass m and momentum p is given by
$$E^2 = m^2c^4 + p^2c^2$$
which reduces to the well-known $E = mc^2$ when the momentum is zero. But this isn't quite the whole story. Because the more familiar equation comes from taking the square root of the full equation, in mathematics we have to say that E can be either positive or negative. Just as

---

*Quantum Theory and Beyond, page 1.

$2 \times 2 = 4$, so does $-2 \times -2 = 4$, and strictly speaking $E = \pm mc^2$. When such "negative roots" crop up in the equations, as often as not they can be dismissed as meaningless, and it is "obvious" that the only answer that we are interested in is the positive root. Dirac, being a genius, did not take this obvious step, but puzzled over the implications. When energy levels are calculated in the relativistic version of quantum mechanics, there are two sets, one all positive, corresponding to mc^2, and the other all negative, corresponding to $-mc^2$. Electrons ought, according to the theory, fall into the lowest unoccupied energy state, and even the highest negative energy state is lower than the lowest positive energy state. So what do the negative energy levels mean, and why didn't all the electrons in the universe fall into them and disappear?

Dirac's answer hinged upon the fact that electrons are fermions, and that only one electron can go into each possible state (two per energy level, one with each spin). It must be, he reasoned, that electrons didn't fall into the negative energy states because all those states are already full. What we call "empty space" is actually a sea of negative energy electrons! And he didn't stop there. Give an electron energy, and it will jump up the ladder of energy states. So, if we give an electron in the negative energy sea enough energy it ought to jump up into the real world and become visible as an ordinary electron. To get from the state $-mc^2$ to the state $+mc^2$ clearly requires an input of energy of $2mc^2$, which, for the mass of an electron, is about 1 MeV and can be provided quite easily in atomic process or when particles collide with one another. The negative energy electron promoted into the real world would be normal in every respect, but it would leave behind a hole in the negative energy sea, the *absence* of a negatively charged electron. Such a hole, said Dirac, ought to behave like a positively charged particle (much as a double negative makes an affirmative, the absence of a negatively charged particle in a negative sea ought to show up as a positive charge). When he first thought of the idea, he reasoned that because of the symmetry of the situation this positively charged particle ought to have the same mass as the electron. But in a moment of weakness when he published the idea he suggested that the positive particle might be the proton, which was the only other particle known in the late 1920s. As he describes in *Directions in Physics,* this was quite wrong, and he should have had the courage to predict that experimenters would find a previously unknown particle with the same mass as the electron but a positive charge.

Nobody was quite sure how to take Dirac's work at first. The idea that the positive counterpart to the electron was the proton was dismissed, but nobody took the idea very seriously, until Carl Anderson, an American physicist, discovered the trace of a positively charged particle during his pioneering observations of cosmic rays in 1932. Cosmic rays are energetic particles that arrive at the earth from space. They had been discovered by the Austrian Victor Hess before the First World War, and he shared the Nobel Prize with Anderson in 1936. Anderson's experiments involved tracking charged particles as they moved through a cloud chamber, a device in which the particles leave a trail like the condensation trail of an aircraft, and he found that some particles produced tracks that were bent by a magnetic field the same amount as the track of an electron, but in the opposite direction. They could only be particles with the same mass as the electron but positive charge, and they were dubbed "positrons." Anderson received a Noble Prize for the discovery in 1936, three years after Dirac had received his own Prize, and the discovery transformed physicists' view of the particle world. They had long suspected the existence of a neutral atomic particle, the neutron, which James Chadwick found in 1932 (for which he received the 1935 Nobel Prize), and they were fairly happy with the idea of an atomic nucleus made up of positive protons and neutral neutrons, surrounded by negative electrons. But positrons had no place in this scheme of things, and the idea that particles could be created out of energy changed the concept of a fundamental particle entirely.

Any particle can, in principle, be produced by the Dirac process from energy, provided that it is always accompanied by the production of its antiparticle counterpart, the "hole" in the negative energy sea. Although physicists prefer more erudite versions of the particle-creation story today, the rules are much the same, and one of the key rules is that whenever a particle meets its antiparticle counterpart it "falls into the hole," liberating energy $2mc^2$ and disappearing, not so much in a puff of smoke as in a burst of gamma rays. Before 1932, many physicists had observed particle tracks in cloud chambers, and many of the tracks they had observed must have been due to positrons, but until Anderson's work it had always been assumed that such tracks were due to electrons moving into an atomic nucleus, rather than positrons moving outward. Physicists were biased against the idea of new particles. Today the situation is reversed and, says Dirac, "People are only too willing to postulate a new particle on the slightest evidence, either theoretical or experimental." (*Directions in Physics,* page 18.) The result is that the particle zoo comprises not just the two fundamental particles known in the 1920s, but more than 200, all of which can be produced by providing sufficient energy in particle accelerators, and most of which are highly unstable, "decaying" very rapidly into a shower of other particles and radiation. Among that zoo, the antiproton and antineutron, discovered in the mid-1950s, are almost lost, but none the less significant confirmation of the correctness of Dirac's original ideas.

Whole books have been written about the particle zoo, and many physicists have built their careers as particle taxonomists. But it seems to me that there cannot be anything very fundamental about such a profusion of particles, and the situation is rather like that in spectroscopy before quantum theory, when spectroscopists could measure and catalogue the relationships between lines in different spectra, but had no idea of the underlying causes of the relationships they observed. Something more truly fundamental must provide the ground rules for the creation of the plethora of known particles, a view that Einstein expressed to his biographer Abraham Pais in the 1950s. "It was apparent that he felt that the time was not ripe to worry about such things and that these particles would eventually ap-

pear as solutions to the equations of a unified field theory."[*]
Thirty years on, it looks very much as if Einstein was right,
and the sketchy outlines of one possible unified theory
which incorporates the particle zoo will be described in the
Epilogue. Here, it is sufficient to point out that the great
explosion of particle physics since the 1940s has its roots in
Dirac's development of quantum theory, the first recipes in
the quantum cookbook.

INSIDE THE NUCLEUS

After the triumphs of quantum mechanics in explaining
the behavior of atoms, it was natural for physicists to turn
their attention to nuclear physics, but in spite of many practical successes (including the reactor at Three Mile Island
and the hydrogen bomb) we still do not have as clear an
idea of what makes the nucleus tick as we do of the behavior of the atom. This isn't really so surprising. In terms
of its radius, the nucleus is 100,000 times smaller than the
atom; since volume is proportional to the cube of radius, it
is more meaningful to say that the atom is a thousand million million (10^{15}) times bigger than the nucleus. Simple
things like mass and charge of the nucleus can be measured, and these measurements lead to the concept of isotopes, nuclei that have the same number of protons, and
therefore form atoms with the same number of electrons
(and the same chemical properties) but different numbers
of neutrons, and therefore different mass.

Since all the protons packed into the nucleus have
positive charge, and therefore repel each other, there must
be some stronger form of "glue" holding them together, a
force that only works across the very short ranges corresponding to the size of the nucleus, and this is called the
strong nuclear force (there is also a weak nuclear force,
which is weaker than the electric force but plays an important part in some nuclear reactions). And it looks as if the

---

Subtle Is the Lord, page 8.

neutrons play a part in the stability of the nucleus as well, because simply by counting the numbers of protons and neutrons in stable nuclei physicists come up with a picture rather like the shell picture of electrons around the nucleus. The largest number of protons found in any naturally occurring nucleus is 92, in uranium. Although physicists have succeeded in manufacturing nuclei with up to 106 protons; these are unstable (except for some isotopes of plutonium, atomic number 94) and break up into other nuclei. Altogether, there are some 260 known stable nuclei; the state of our knowledge about those nuclei, even today, is rather less adequate than the Bohr model is as a description of the atom, but there are clear signs of some sort of structure within the nucleus.

Nuclei that have 2, 8, 20, 28, 50, 82, and 126 nucleons (neutrons or protons) are particularly stable, and the corresponding elements are much more abundant in nature than elements corresponding to atoms with slightly different numbers of nucleons, so these are sometimes called "magic numbers." But protons dominate the structure of the nucleus, and for each element there is only a limited range of possible isotopes corresponding to different numbers of neutrons—the possible number of neutrons is generally a little bigger than the number of protons, and gets bigger for heavier elements. Nuclei that possess magic numbers of *both* protons and neutrons are particularly stable, and theorists predict on this basis that superheavy elements with about 114 protons and 184 neutrons in their nuclei ought to be stable—but these massive nuclei have never been found in nature or made in particle accelerators by sticking more nucleons onto the most massive nuclei that occur in nature.

The most stable nucleus of all is iron-56, and lighter nuclei would "like" to gain nucleons and become iron, while heavier nuclei would "like" to lose nucleons and move toward the most stable form. Inside stars, the lightest nuclei, hydrogen and helium, are converted into heavier nuclei in a series of nuclear reactions that fuse the light nuclei together, making elements such as carbon and oxygen along the road to iron, and releasing energy as a result.

When some stars explode as supernovae, a great deal of gravitational energy is put into the nuclear processes, and this pushes the fusion beyond iron to make heavier elements, including things like uranium and plutonium. When heavy elements move back toward the most stable configuration, by ejecting nucleons in the form of alpha particles, electrons, positrons, or individual neutrons, they too release energy, which is essentially the stored-up energy of a long-gone supernova explosion. An alpha particle is essentially the nucleus of a helium atom and contains two protons and two neutrons. By ejecting such a particle, a nucleus reduces its mass by four units, and its atomic number by two. And it does so in accordance with the rules of quantum mechanics and the uncertainty relations discovered by Heisenberg.

The nucleons are held together inside the nucleus by the strong nuclear force, but if an alpha particle was just outside the nucleus it would be strongly repelled by the electric force. The combined effect of the two forces is to make what physicists call a "potential well." Imagine a cross-section through a volcano with gently sloping sides and a deep crater. A ball placed just outside the crater rim

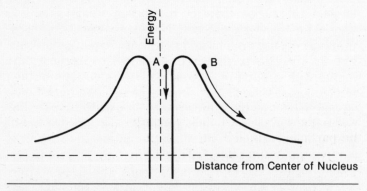

*Figure 7.1/*A potental well at the heart of an atomic nucleus. A particle at A has to stay inside the well unless it can gain enough energy to jump "over the top" to B, when it will rush away "downhill." Quantum uncertainty allows a particle, occasionally, to "tunnel" through from A to B (or B to A) without having enough energy of its own to climb the hill.

will roll away down the outside of the mountain; one placed just inside the crater rim will fall into the heart of the volcano. Nucleons inside the nucleus are in a similar situation—they are inside the well at the heart of the atom, but if they could just get over the "rim"—even by a tiny amount—they would "roll away," pushed by the electric force. The snag is, according to classical mechanics nucleons, or groups of nucleons such as an alpha particle, just don't have enough energy to climb out of the well and over the rim—if they did, they wouldn't be in the well in the first place. The quantum mechanical view of the situation, however, is rather different. Although the potential well still provides a barrier, it is not insurmountable, and there is a definite, if small, probability that an alpha particle might actually be outside, not inside, the nucleus. In terms of uncertainty, one of the Heisenberg relations involves energy and time, and says that any particle's energy can only be defined within a range ΔE over a period of time Δt, such that $\Delta E \times \Delta t$ is bigger than $\hbar$. For a short time, a particle can "borrow" energy from the uncertainty relation, gaining enough energy to jump over the potential barrier before giving it back. When it returns to its "proper" energy state, it is just outside the barrier instead of just inside, and rushes away.

Or you can look at it in terms of the uncertainty of position. A particle that "belongs" just inside the barrier may appear just outside, because its position is only fuzzily determined in quantum mechanics. The larger the energy of the particle, the easier it is for it to escape, but it does *not* have to have enough energy to climb out of the potential well in the way classical theory requires. The process is as if the particle tunneled out through the barrier, and it is purely a quantum effect.* This is the basis of radioactive

---

*The same process operates in reverse when nuclei fuse together. When two light nuclei are pushed together by the pressure inside a star, they can only fuse if they overcome the potential barrier from the outside. The amount of energy each nucleus possesses in that situation depends on the temperature at the heart of the star, and in the 1920s astrophysicists were puzzled to find that the calculated temperature inside the sun is a little less than it ought to be—the nuclei at the heart of the sun do not have enough energy to overcome

decay; but to explain nuclear fission we have to turn to a
different model of the nucleus.

Forget about the individual nucleons in their shells for
the time being, and consider the nucleus as a droplet of
liquid. Just as a drop of water wobbles in a changing pattern of shapes, so some of the collective properties of the
nucleus can be explained as due to the changing shape of
the nucleus. A large nucleus can be thought of as wobbling
in and out, changing shape from a sphere to something like
a fat dumbbell and back again. If energy is put into such a
nucleus, the oscillation may become so extreme that it
breaks the nucleus in two, splitting off two smaller nuclei
and a spattering of tiny droplets, alpha and beta particles
and neutrons. For some nuclei, this splitting can be triggered by the collision of a fast-moving neutron with the
nucleus, and a chain reaction occurs when each nucleus
fissioned in this way produces enough neutrons to ensure
that at least two more nuclei in its neighborhood also fission. For uranium-235, which contains 92 protons and 143
neutrons, two unequal nuclei with atomic numbers in the
range 34 to 58, and adding up to 92, are always produced,
with a scattering of free neutrons. Each fission releases
about 200 MeV of energy, and each one sets off several
more, provided that the lump of uranium is big enough that
the neutrons do not escape from it altogether. Left to run
away exponentially, this is the process of the atomic bomb;
moderated by using a material that absorbs neutrons to
keep the process just ticking over, we have a controlled fission reactor that can be used to heat water into steam and
generate electricity. Once again, the energy we extract is
the stored energy of a stellar explosion, long ago and far
away.

In the fusion process, however, we can mimic the energy production of a star like the sun here on earth. So far,

---

the potential barrier and fuse together, according to classical mechanics. The
answer is that some of them tunnel through the barrier at a slightly lower
energy, in line with the rules of quantum mechanics. Among other things,
quantum theory explains why the sun shines, when classical theory says it
cannot.

we have only been able to copy the first step up the fusion ladder, from hydrogen to helium, and we have not been able to control the reaction, only to let it run away in the hydrogen, or fusion, bomb. The trick with fusion is the opposite of the trick with fission. Instead of encouraging a large nucleus to break up, you have to force small nuclei together, against the natural electrostatic repulsion of their positive charges, until they are so close that the strong nuclear force, which only has a very short range, can overwhelm the electric force and pull them together. As soon as you get a few nuclei to fuse in this way, the heat generated in the process causes an outward rush of energy that tends to blow apart any other nuclei on the point of fusing, and stops the whole process in its tracks.* The hope of unlimited energy for the future from nuclear fusion depends on finding a way to hold enough nuclei together in one place for long enough to get a useful amount of energy out. It is also rather important to find a process that releases more energy than we have used pushing the nuclei together in the first place. It's easy enough in a bomb—essentially, you just surround the nuclei you want to fuse with uranium, then trigger the uranium into a fission explosion. The inward pressure from the surrounding explosion will then bring enough hydrogen nuclei into contact to set off the second, and more spectacular, fusion explosion. But something a little more subtle is required for civil power stations, and techniques now being investigated include the use of strong magnetic fields shaped to act as a bottle holding in the charged nuclei, and pulses of light from laser beams which physically squeeze the nuclei to-

---

*One way to gain energy from fusion is by combining an isotope of hydrogen, which has one proton and one neutron (deuterium), with one which has one proton and two neutrons (tritium). The result is a helium nucleus (two protons, two neutrons), a free neutron, and 17.6 MeV of energy. Stars work by more complicated processes involving nuclear reactions between hydrogen and nuclei such as carbon that are present in small quantities inside the star. The net effect of such reactions is to fuse four protons into a helium nucleus, with two electrons and 26.7 MeV of energy being released, and the carbon put back into circulation to catalyze another cycle of reactions. But it is processes involving tritium and deuterium that are being investigated in fusion laboratories here on earth.

gether. Lasers, of course, are manufactured in accordance with another recipe from the quantum cookbook.

LASERS AND MASERS

Although it took a master chef like Dirac to discover the recipes for making new particles in quantum cookery, nuclear processes are understood in terms that are less complete than the Bohr model of the atom. So, perhaps, it should not be too much of a surprise to find that the Bohr model itself still has its uses. Some of the most exotic and exciting of recent scientific developments, the lasers, can be understood by any competent quantum short-order cook who has heard of the Bohr model, and require no great genius in their interpretation. (The genius comes in, in this case, in the technology of their construction, but that is another story.) So, with apologies to Heisenberg, Born, Jordan and Dirac, and Schrödinger, let's ignore all the quantum subtleties for a while and go back to that neat model of electrons orbiting around the nucleus of an atom. Remember that when an atom gains a quantum of energy an electron jumps up to a different orbit, on this picture, and that when such an excited atom is left alone then, sooner or later, the electron will fall back to the ground state, releasing a very precisely defined quantum of radiation with a definite wavelength. The process is called spontaneous emission and is the counterpart of absorption.

When Einstein was investigating such processes in 1916 and laying the statistical ground rules for quantum theory, which he later found so abhorrent, he realized that there is another possibility. An excited atom can be *triggered* into releasing its extra energy and going back into the ground state, if it is nudged, as it were, by a passing photon. This process is called stimulated emission, and it only happens if the passing photon has exactly the same wavelength as the one that the atom is primed to radiate. Rather like the cascade of neutrons that is involved in a chain reaction of nuclear fission, we can imagine an array of excited atoms with just one photon of the right wavelength coming along and stimulating one atom to radiate; the original photon plus the new one then stimulate two more atoms to radiate, the four photons together trigger four more, and so on. The result is a cascade of radiation, all with precisely the same frequency. Furthermore, because of the way the emission is triggered, all of the waves move precisely in step with one another—all the waves go "up" together, and all the troughs go "down" together, producing a very pure beam of what is called coherent radiation. Because none of the peaks and troughs in such radiation are canceling each other out, all of the energy released by the atoms is present in the beam and can be delivered onto a small area of material that the beam is shone upon.

When a collection of atoms or molecules is excited by heat, they fill up a band of energy levels and, left to their own devices, radiate different wavelengths of energy in an incoherent and jumbled fashion, carrying much less effective energy than the atoms and molecules release. But there are tricks that can be used to fill up a narrow band of energy levels preferentially, and then to trigger the return of the excited atoms in this band to their ground state. The trigger for the cascade is a weak input of radiation of the right frequency; the output is a much stronger, amplified beam with the same frequency. The techniques were first developed in the late 1940s, independently by teams in the U.S. and the U.S.S.R., using radiation in the radio band of the spectrum from about 1 cm to 30 cm, called the microwave band; the pioneers received the Nobel Prize for their work in 1954. Because the radiation in this band is called microwave radiation, and because the process involves amplification of microwaves by the stimulated emission of radiation in line with Einstein's ideas of 1917, the pioneers coined the name microwave amplification by stimulated emission of radiation, and the acronym MASER, for the process.

It was ten years before anyone succeeded in finding a way to make the trick work for optical frequencies of radiation, and then in 1957 two people hit on the same idea more or less simultaneously. One (who seems to have been first) was Gordon Gould, a graduate student at Columbia University; the other was Charles Townes, one of the maser pioneers who had shared the Nobel Prize in 1964. The arguments about exactly who discovered what and when have been subject to a legal battle about patent rights, since lasers, the optical equivalent of masers (from "light amplification . . .") are now big business and big money, but happily we don't have to get embroiled in that issue. Today, there are several different kinds of laser, the simplest being the optically pumped solid lasers.

In this design, a rod of material (such as ruby) is prepared with polished, flat ends and surrounded by a bright light source, a gas discharge tube that can flash rapidly on and off, producing pulses of light with sufficient energy to excite the atoms in the rod. The whole apparatus is kept cool to ensure the minimum amount of interference from thermal excitation of the atoms in the rod, and the bright flashes from the lamp are used to stimulate (or pump) the atoms into an excited state. When the laser is triggered, a pulse of pure ruby light, carrying thousands of watts of energy, emerges from the flat end of the rod.

Variations on the theme include liquid lasers, fluorescent-dye lasers, gas lasers, and so on. All share the same essential features—incoherent energy is put in, and coherent light comes out in a pure pulse carrying a lot of energy. Some, like the gas lasers, give a continuous, pure beam of light that is the ultimate "straight edge" for surveying purposes, and that has found widespread use at rock concerts and in advertising. Others produce short-lived but powerful pulses of energy that can be used to drill holes in hard objects (and might one day have military applications). Laser cutting tools are used in situations as diverse as the clothing industry and microsurgery. And laser beams can be used to carry information far more effectively than radio waves, since the amount of information that can be passed each second increases as the frequency of the radiation used increases. The bar codes on many supermarket products (and on the cover of this book) are read by a laser scanner; the video disks and compact audio disks that came on the market in the early 1980s are scanned by lasers; genuine three-dimensional photographs, or holograms, can be made with the aid of lasers; and so on.

The list is virtually endless, even before we include the applications of masers in amplifying faint signals (for example, from communications satellites), radar and elsewhere; and it all stems not even from quantum theory proper, but from the first version of quantum physics. When you buy a packet of cornflakes and have it scanned by a laser at the checkout, or attend a rock concert with spectacular colored-laser displays, or watch the concert on TV though a satellite link halfway around the world, or listen to the latest recording of that same band on the latest hi-fi compact disk system, or admire the magic of a holographic reproduction, it is all thanks to Albert Einstein and Niels Bohr, who laid down the principles of stimulated emission more than sixty years ago.

THE MIGHTY MICRO

The most pervasive influence of quantum mechanics on our everyday lives is undoubtedly in the area of solid-state physics. The name "solid state" itself is unromantic; even if you've heard it, you probably don't associate it with quantum theory. Yet it is the branch of physics that has given us the transistor radio, the Sony Walkman, digital watches, pocket calculators, micro computers, and programmable washing machines. Ignorance about solid-state physics isn't because it is an esoteric branch of science, but because it is so commonplace that it is taken for granted. And, once again, we would have none of these devices without a satisfactory grasp of quantum cookery.

All of the devices mentioned in the paragraph above depend on the properties of semiconductors, which are solids, that, logically enough, have properties intermediate between those of conductors and insulators. Without going into details, insulators are substances that do not conduct electricity, and they do not conduct because the electrons in their atoms are firmly bound to the nuclei, in line with the rules of quantum mechanics. In conductors, such as metals, it just happens that each atom has some electrons that are only loosely bound to the nucleus, and are in energy states up near the top of the atomic potential well. When the atoms are grouped together in a solid, the top of one energy well blends into the well belonging to the next atom along, and electrons in these high levels are free to wander along from one atomic nucleus to the next, no longer really bound to any one nucleus, and able to carry an electric current through the metal.

Ultimately, the property of conduction depends on the Fermi-Dirac statistics, which forbid these loosely bound electrons to fall down deep into the atomic potential wells, where the energy states for tightly bound electrons are all fully occupied. If you try to squeeze a metal, it resists the pressure; metals are strong. And the reason why metals are so strong, so resistant to pressure, is that because of the Pauli exclusion principle for fermions the electrons cannot be squeezed tighter together.

The energy levels for electrons in a solid are calculated using the quantum mechanical wave equations. Electrons that are bound tightly to the nuclei are said to be in the valence band of a solid, and electrons that are free to wander from nucleus to nucleus are said to be in the conduction band. In an insulator, all the electrons are in the valence band; in a conductor, some have overflowed into the conduction band.* In a semiconductor, the valence band is full, and there is only a narrow energy gap between this band and the conduction band—typically about 1 eV. So it is easy for an electron to hop into the conduction band and carry an electric current through the material. Unlike the situation for a conductor, though, this electron that has gained energy leaves a hole behind in the valence band. In exactly the same way that Dirac reasoned for the creation of electrons and positrons out of energy, this absence of a

---

*There is actually another type of conductor in which the valence band itself is not filled, so that electrons can move about within the valence band.

negatively charged electron in the valence band behaves, as far as electrical properties are concerned, like a positive charge. So a natural semiconductor typically has a few electrons in its conduction band, and a few positively charged holes in its valence band, both of which can convey electric current. You can think of successive electrons falling into the hole in the valence band and leaving a hole behind them, into which the next electron hops, and so on; or you can think of the holes as real, positive particles moving in the opposite direction. As far as electric currents are concerned, the effect is the same.

Natural semiconductors would be interesting enough, not least because of the neat analogy they provide for the creation of an electron-positron pair. But it is very difficult to control their electrical properties, and it is control that has made these materials so important to our everyday lives. The control is achieved by creating artificial semiconductors, one sort dominated by free electrons, the other by free "holes."

Once again, the trick is easy to understand, not so easy to work in practice. In a crystal of germanium, for example, each atom has four electrons in its outer shell (this is short-order quantum cookery, and the Bohr model is adequate for the job) which are "shared" with neighboring atoms to make the chemical bonds that hold the crystal together. If the germanium is "doped" with a few atoms of arsenic, the germanium atoms still dominate the structure of the crystal lattice, and the arsenic atoms have to squeeze in as best they can. In chemical terms, the main difference between arsenic and germanium is that arsenic has a fifth electron in its outer shell, and the best way for an arsenic atom to squeeze into a germanium lattice is to discard the extra electron and take up four chemical bonds, pretending it is a germanium atom. The extra electrons provided by the arsenic atoms wander through the conduction band of the semiconductor thus created, and there are no corresponding holes. Such a crystal is called an n-type semiconductor.

The alternative is to dope the germanium (sticking to our original example) with gallium, which has only three electrons available for chemical bonding. The effect is as if we create a hole in the valence band for every atom of gallium present, and the valence electrons move by jumping into the holes, which behave like positive charges. Such a crystal is called a p-type semiconductor. And things get interesting when the two types of semiconductor are put in contact with one another. An excess of positive charge on one side of the barrier, and negative charge on the other, produces an electric potential difference that tries to push electrons in one direction and opposes their motion in the other direction; such a joined pair of semiconducting crystals, called a diode, effectively only allows electric current to pass in one direction. Slightly more subtly, electrons can be encouraged to jump the gap from n to p and into a hole, emitting a spark of light as they do so. Diodes designed to produce light in this fashion are called light emitting diodes, or LEDs, and are used for the numeric displays in some pocket calculators and watches, and other visual displays. A diode that works in the opposite direction, absorbing light and pumping an electron out of a hole and into the neighboring conduction band is a photodiode, used to ensure that an electric current only flows when a beam of light is shining on the semiconductor. This is the basis for automatic door-opening devices that do their job when you move in front of the light beam. But there is more to semiconductors than diodes.

When three pieces of semiconductor are put together to make a sandwich (pnp, or npn), the result is a transistor (each piece of the transistor is usually connected into an electric circuit, so transistors in your radio, say, can be identified by the three spidery legs sticking out of the metal or plastic can that houses the semiconductor itself). With suitably doped materials, it is possible to set up an arrangement where a small flow of electrons across one np junction causes a much larger flow across the other junction in the sandwich—the transistor acts as an amplifier. As any electronics buff knows, the two components of diode and amplifier are the key to designing a sound system. But even transistors are pretty old hat today, and you won't find any three-legged cans in your radio unless it is an old "tranny."

Up until the fifties, we relied upon the cumbersome old

"wireless" for entertainment, an apparatus which, in spite of its name, was stuffed full of wires and glowing vacuum tubes that did the same job that semiconductors now do. By the end of the fifties, the transistor revolution was under way, and the large, glowing valves were replaced by transistors while the wires were replaced by boards onto which the circuitry was printed and the transistors were soldered. From this it was a short step to the integrated circuit, where all the circuits and the semiconductor amplifiers, diodes, and so on were made together in one piece and simply plugged together to make the heart of a radio, cassette player, or whatever; and at the same time a similar revolution was taking place in the computer industry.

Like the old wirelesses, the first computers were big and cumbersome. They were full of valves and contained miles of wiring. Even twenty years ago, with the first solid-state revolution in full swing, a computer to do the same work as a modern micro the size of a typewriter would have required the ground floor of a house to accommodate its "brain" and more space for its associated air-conditioning plant. The revolution that has put that sort of computing power onto a desk-top machine costing a few hundred dollars is the same one that has turned grandpa's tabletop wireless into a radio the size of a packet of cigarettes, and takes the solid-state revolution from the transistor to the chip.

Biological brains and electronic computers are both in the business of switching. Your brain contains about 10,000 million switches in the form of neurons, made from nerve cells; a computer has switches made of diodes and transistors. In 1950 a computer with the same number of switches as your brain would have been as big as the island of Manhattan; today, by sticking micro chips together, it might just be possible to pack as many switches into the volume of a human brain, although wiring such a computer up would be a problem and it hasn't yet been done. But the example indicates just how small the chip is, even compared with the transistor.

The semiconductor used in a standard micro chip today is silicate—basically nothing more than common sand

Given the right encouragement, electricity will flow through silicate; without encouragement it will not. Long crystals of silicate about 10 cm across are cut into razor-thin wafers and chopped into hundreds of small rectangular chips, each smaller than a match head, and onto each chip is pressed, layer upon layer like some delicate Greek pastry, a dense complex of fine electronic circuitry, the equivalent of transistors, diodes, integrated circuits and all. One chip is, effectively, a whole computer, and all the rest of the works of a modern micro are concerned with getting information in and out of the chip. And they are so cheap to make (once the substantial costs of designing the circuitry and setting up the machines to reproduce it have been met) that they can be turned out in the hundreds, tested, and the ones that don't work simply thrown away. To make *one* chip, starting from scratch, might cost a million dollars; to make as many more as you like the same as the first would then cost a few pennies each.

So there are a few more things in the everyday world to lay at the door of the quantum. Recipes from just one chapter in the quantum cookbook have given us digital watches, home computers, the electronic brains that guide the space shuttle into orbit (and sometimes decide not to let it fly, whatever the human operators may say), portable TV, personal stereo systems and powerful hi-fi which can deafen you, and better deaf aids to compensate for the resulting hearing loss. Genuinely portable (pocket-sized) computers can't be far off; genuinely intelligent machines are a more remote, but realistic, possibility. The computers that control Mars landers and the *Voyager* probes to the outer solar system are first cousins to the chips that control the arcade games, and all have their roots in the strange behavior of electrons in obedience with the basic quantum rules. Even the story of the mighty micro, however, doesn't exhaust the potential of solid-state physics.

SUPERCONDUCTORS

Like semiconductors, superconductors have a logical name. A superconductor is a material that conducts electricity without any apparent resistance at all. This is as close as we are ever likely to get to perpetual motion—it isn't quite something for nothing, but it is a rare example of actually getting everything you pay for in physics, without being short changed. And it can be explained by a change that makes pairs of electrons associate with one another and move together. Although each electron has half-integer spin, and therefore obeys Fermi-Dirac statistics and the exclusion principle, a pair of electrons can behave in some circumstances as a single particle with integer spin. Such a particle is not constrained by the exclusion principle, and obeys the same Bose-Einstein statistics that describe, in quantum mechanical terms, the behavior of photons.

The Dutch physicist Kamerlingh Onnes discovered superconductivity in 1911, when he found that mercury lost all its electrical resistance when cooled to below 4.2 degrees on the absolute temperature scale (4.2 degrees K, or about −269 degrees C). Onnes got the Nobel Prize for his low-temperature work in 1913, but this was for other work, notably the preparation of liquid helium, and the phenomenon of superconductivity wasn't satisfactorily explained until 1957, when John Bardeen, Leon Cooper, and Robert Schrieffer came up with a theory that earned them the Nobel Prize in Physics in 1972.* The explanation depends on the way paired electrons interact with atoms in a crystal lattice. One electron interacts with the crystal, and as a result of this interaction the interaction of the crystal with the other electron in the pair is modified. So, in spite of their natural tendency to repel one another, the pair of elec-

---

*Bardeen had already made a name for himself in 1948 for his work with William Shockley and Walter Brattain on an invention that got the three of them the Nobel Prize in 1956. This little invention was the transistor, and Bardeen is the first person to have won the physics prize twice.

trons forms a loosely bound association, sufficient to account for the change from Fermi-Dirac to Bose-Einstein statistics. Not all materials can become superconducters, and even in those that can any small disturbance from the thermal vibrations of atoms in the crystal will break up the electron pairing, which is why the phenomenon only occurs at very low temperatures, in the range 1 to 10K. Below some critical temperature, which varies from one material to another but is always the same for the same substance, some materials become superconductors; above that temperature, the electron pairing is broken and they have normal electrical properties.

The theory is borne out by the fact that materials that are good conductors at room temperature are not the best superconductors. A good "normal" conductor allows electrons to move freely precisely because they do not interact very much with the atoms in the crystal lattice—yet without an interaction between the electrons and the atoms there is no way for the electron coupling that leads to superconductivity to be effective at low temperatures.

It is unfortunate that superconductors have to be made

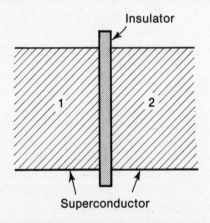

Figure 7.2/Strange things happen at a Josephson junction, when two pieces of superconductor are separated by a layer of insulation. Under the right circumstances, electrons can tunnel through the barrier.

so cold before they perform, because the potential uses for a more convenient superconductor are easy to imagine—transmission of power along cables without any loss of energy being simply the most obvious example. Superconductors also perform other tricks. A normally conducting metal can be penetrated by a magnetic field, but a superconductor sets up electric currents on its surface that repel and expel the magnetic field—the perfect screen against unwanted interference from magnetic fields, but impractical as long as the screen has to be cooled to a few degrees K. When two superconductors are separated by an insulator, you might expect no current to flow; but remember that the electron is obeying the same quantum rules that allow particles to tunnel out of the nucleus. If the barrier is thin enough, the probability of electron pairs being able to cross the gap is significant, but it doesn't produce commonsense results. Such junctions (called Josephson junctions) produce *no* current if there is a potential difference across the barrier, but there *is* a current if the voltage from one side to the other is zero. And a double Josephson junction, made by taking two pieces of superconductor shaped like tuning

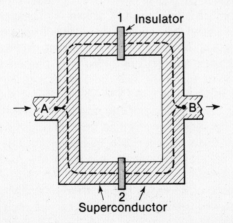

*Figure 7.3/*Two Josephson junctions can be arranged to make a system analogous to the two-slit experiment for light. With this setup, interference between electrons can be observed, one of many indications of the wave nature of these "particles."

forks and pressing the double ends together, separated by a sandwiched layer of insulator, can be made to mimic the quantum mechanical behavior of the electron "double slit" experiment, which we shall deal with in detail in the next chapter, and which is the cornerstone of some of the strangest features of the quantum world.

Not only electrons can join together to make pseudo-bosons that defy the everyday laws of physics at low temperatures. Helium atoms can do a very similar trick, and this is the basis of a property of liquid helium called super-fluidity. When you stir a cup of coffee and then leave it alone, the spinning swirl of liquid slows and then stops because of the fluid equivalent of friction or viscous forces. Try the same thing with helium cooled below 2.17K and the spinning will never stop; even left entirely to its own devices, the fluid may crawl up the side of a bowl and over the top, and instead of finding it difficult to pass through a narrow tube, superfluid helium flows more easily the narrower the tube in which it is confined. All of this strange behavior can be explained in terms of Bose-Einstein statistics, and although once again the low temperatures required make it difficult to find practical uses for the phenomenon, the behavior of atoms at these low temperatures, like the behavior of electrons in superconductivity, provides an opportunity to see quantum processes in action. If a little superfluid helium is placed in a tiny bucket, 2 mm or so across, and the bucket is rotated, at first the helium stays at rest. As the speed of the rotation increases, at a critical value of the angular momentum all of the helium develops an angular flow, changing from one quantum state to another. No in-between state—corresponding to in-between angular momentum—is allowed by the quantum rules, and the whole collection of helium atoms, a visible mass vastly bigger than an individual atom or the particles of the quantum world, can be seen acting in accordance with the quantum rules. Superconductivity, as we shall see later, can also be applied to objects on the human, rather than the atomic, scale. But quantum theory is not restricted to the world of physics, or even the physical sciences. All of chemistry, remember, is now understood in terms of the quantum ground rules. Chemistry is the science of molecules, rather than individual atoms and subunits of atoms, and that includes the most important molecules for all of us—living molecules, including the molecule of life, DNA. Our present understanding of life itself is firmly rooted in quantum theory.

LIFE ITSELF

Quite apart from the scientific importance of quantum theory to an understanding of the chemistry of life, there are direct personal links between some of the leading characters in the quantum story and the discovery of the double helix structure of DNA, the molecule of life. The laws that describe the diffraction of X rays from crystals were discovered by Lawrence Bragg and his father, William, working at the Cavendish in the years before the First World War; they jointly received the Nobel Prize for this work, Lawrence at such an early age (in 1915, when he was a serving officer in France) that he was still alive (in spite of having served in France in the First World War) to celebrate the golden jubilee of the occasion fifty years later. The older Bragg had made his initial reputation in physics with studies of alpha, beta, and gamma radiation, and in the late years of the first decade of the twentieth century he had shown that both gamma and X rays behave in some respects like particles. Bragg's law of X-ray diffraction, which is the key to unlocking the secrets of the structures of crystals depends, however, on the wave properties of X rays that are bounced off the atoms in a crystal. The interference patterns produced as a result depend on the spacing of the atoms in the crystal and the wavelength of the X rays, and in skilled hands this tool has been developed to pinpoint the positions of individual atoms even in very complex crystal structures.

The insight that led to Bragg's Law came in 1912, chiefly from Lawrence Bragg; by the late 1930s he was Cavendish Professor of Physics in Cambridge (succeeding

Rutherford on his death in 1937) and still actively involved in X-ray work, among many other things. It was during this decade that the new science of biophysics began to make progress. The pioneering work of J. D. Bernal in determining the structure and composition of biological molecules by X-ray diffraction led to detailed investigations of the complex protein molecules that carry out many of the functions of life. Researchers Max Perutz and John Kendrew shared the Nobel Prize in Chemistry in 1962 for determining the structures of hemoglobin (the molecule that carries oxygen in your blood) and myoglobin (a muscle protein) as a result of research that started in Cambridge before the Second World War.

The names that are forever linked in popular mythology with the origins of molecular biology are, however, those of the "young Turks" Francis Crick and James Watson, who developed the double-helix model of DNA in the early 1950s, and received the Nobel Prize "for Physiology or Medicine" (jointly with Maurice Wilkins), also in 1962. The flexibility of the Nobel Committee in managing to honor different pioneers in the field of biophysics by awarding prizes in the same year under the headings "chemistry" and "physiology" is admirable, but it is unfortunate that the strict rules against posthumous awards prevented them giving a share of the Crick-Watson-Wilkins prize to Wilkins' colleague Rosalind Franklin, who had done much of the key crystalographic work that revealed the structure of DNA, but who died in 1958 at the age of thirty-seven. Franklin's place in popular mythology is as the fire-breathing feminist of Watson's book *The Double Helix,* a colorful personal account of his time in Cambridge that is highly entertaining but far from being a fair and accurate portrayal of his colleagues, or even of himself.

The work that led Watson and Crick to the structure of DNA was carried out at the Cavendish, where Bragg still reigned. Watson, a young American in Europe to do postdoctoral research, describes in his book how he first encountered Bragg when seeking permission to work at the Cavendish. The white-moustached figure, now in his early sixties, struck young Watson as a relic of the scientific past, undoubtedly one who now spent most of his days sitting in London clubs. But permission was granted, and Watson was surprised by the active interest Bragg took in research, providing invaluable, though not always welcome, guidance on the path to the solution of the DNA problem. Francis Crick, although older than Watson, was technically still a student, working for his PhD. His scientific career, like that of many others of his generation, had been interrupted by the Second World War, although in his case this may have been no bad thing. He originally trained as a physicist, and it was only in the late 1940s that he moved toward the biological sciences, a decision that was inspired in no small measure by a little book written by Schrödinger and published in 1944. The book, titled *What Is Life?*, is a classic— still in print, and well worth seeking out—that expounded the idea that the fundamental molecules of life could be understood in terms of the laws of physics. The important molecules to explain in those terms are the genes that carry information about how a living body is to be constructed and how it is to operate. When Schrödinger wrote *What Is Life?* it was thought that genes, like so many other living molecules, were made of protein; just about at that time, however, it was being discovered that hereditary traits are actually carried by molecules of an acid called deoxyribonucleic acid, found in the central nuclei of living cells.* This is DNA, and it is the structure of DNA that Crick and Watson determined, using the X-ray data obtained by Wilkins and Franklin.

I have described the detailed structure of DNA and its role in the life process in another book.† The key feature is that DNA is a double molecule, made of two strands twisted around one another. The order in which different chemical components, called bases, are strung along the DNA spines carries information that the living cell uses to construct the protein molecules that do all the work, like carrying oxygen around the blood or making muscles operate. A strand of

---

*The original use of the same term "nucleus" for the central part of an atom was a deliberate echo of the already existing biological terminology.
†*The Monkey Puzzle*, with Jeremy Cherfas.

DNA can unravel partially, revealing a string of bases that act as a template for the construction of the other molecules, or it can untwist completely and replicate itself by matching up every base along the spine of the strand with its counterpart and building a mirror image strand to form a new double helix. Both processes use as raw materials the chemical soup inside the living cell; both are essential to life. And mankind is now able to tinker with the coded message along the DNA, altering the instructions coded in the blueprint of life—at least, in the case of some relatively simple living organisms.

This is the basis of genetic engineering. Pieces of genetic material—DNA—can be created by a combination of chemical and biological techniques, and microorganisms such as bacteria can be encouraged to take up this DNA from the chemical soup of their surroundings and incorporate it into their own genetic code. If a strain of bacteria is given the coded information on how to make human insulin in this way, its own biological factories will do just that, producing exactly the material required by diabetics to enable them to lead normal lives. The dream of altering human genetic material to remove the defects that cause problems such as diabetes in the first place is still far from being realized, but there is no theoretical reason why it should not be achieved. A more immediate step, however, will be to use genetic engineering techniques on other animals and plants, producing superior strains for food and other human requirements.

Again, the details can be found elsewhere.* The important point is that we have all heard of genetic engineering and read about the miraculous prospects—and the dangers—it holds for the future. Very few people appreciate, though, that the understanding of living molecules that makes genetic engineering possible depends on our present understanding of quantum mechanics, without which we would not be able to interpret X-ray diffraction data, apart from anything else. To understand how to construct,

---

*For example, in *Man Made Life*, by Jeremy Cherfas.

or reconstruct, genes, we have to understand how and why atoms join together only in certain arrangements, certain distances apart and with chemical bonds of a certain strength. That understanding is the gift of quantum physics to chemistry and to molecular biology.

I've labored the point a little more than I might have had it not been for a member of the University College of Wales. In March 1983, in a review in *New Scientist,* I mentioned in passing that "without quantum theory there would be no genetic engineering, no solid-state computers, no nuclear power stations (or bombs)." This drew a complaint from a correspondent in that respected academic institution to the effect that he was fed up with seeing genetic engineering dragged in everywhere as the new scientific buzzword, and that John Gribbin shouldn't be allowed to get away with such outrageous remarks. What possible connection, however tenuous, could there be between quantum theory and genetics? I hope the connection is clear this time. At one level, it is delightful to be able to point out the fact that Crick's conversion to biophysics was directly inspired by Schrödinger, and that the work that led to the discovery of the DNA double helix was carried out under the formal, if sometimes unwelcome, direction of Lawrence Bragg; at a deeper level, of course, the reason for the interest of pioneers like Bragg and Schrödinger, and the next generation of physicists such as Kendrew, Perutz, Wilkins, and Franklin in biological problems is that these problems are, as Schrödinger pointed out, simply another kind of physics, one that deals with collections of large numbers of atoms in complex molecules.

Far from backing away from that passing comment I made in *New Scientist,* I would strengthen it. If you asked an intelligent, well-read but nonscientific person to summarize the most important contributions of science to our present lives, and to suggest the possible benefits or hazards of scientific progress in the near future, you would surely be given a list that included computer technology (automation, unemployment, entertainment, robots), nuclear power (the bomb, cruise missiles, power stations, Three Mile Island), genetic engineering (new drugs, cloning, the threat of man-made diseases, improved crop strains), and lasers (holography, death rays, microsurgery, communications). Probably the vast majority of such people questioned will have heard of relativity theory, which plays no part in their daily lives; and scarcely any of them will realize that every item on that list has its roots in quantum mechanics, a branch of science that they may never have heard of and almost certainly do not understand.

They are not alone. All of those advances have been achieved by quantum cookery, using the rules that seem to work although no one really understands why they work. In spite of the achievements of the past six decades, it is doubtful whether *anybody* understands *why* the quantum recipes work. The rest of this book will be devoted to probing some of the deeper mysteries that are so often swept under the carpet, and to looking at some of the possibilities and paradoxes.

PART THREE

. . . And Beyond

"It is better to debate a question
without settling it than to
settle a question without debating it."

JOSEPH JOUBERT
1754–1824

CHAPTER EIGHT

CHANCE AND UNCERTAINTY

Heisenberg's uncertainty principle is seen today as a central feature—perhaps *the* central feature—of quantum theory. It wasn't picked up immediately by his colleagues, but took nearly ten years to achieve this exalted position. Since the 1930s, however, its position may have been a little too exalted.

The concept grew out of Schrödinger's visit to Copenhagen in September 1926, the occasion of his famous remark to Bohr about "damned quantum jumping." Heisenberg realized that one of the main reasons why Bohr and Schrödinger sometimes seemed to be at loggerheads was a conflict of concepts. Ideas like "position" and "velocity" (or "spin," which came later) simply don't have the same meaning in the world of microphysics as they do in the everyday world. So what meaning do they have, and how can the two worlds be related? Heisenberg went back to the fundamental equation of quantum mechanics,

$$\mathbf{pq} - \mathbf{qp} = \hbar/i$$

and showed from this that the product of the uncertainties

in position (Δq) and momentum (Δp) must always be bigger than $\hbar$. The same uncertainty rule applies to any pair of what are called conjugate variables, variables that multiply out to have the units of action, like $\hbar$; the units of action are energy × time, and the other most important pair of such variables is indeed energy (E) and time (t). The classical concepts of the everyday world still existed in the microworld, said Heisenberg, but they could only be applied in the restricted sense revealed by the uncertainty relations. The more accurately we know the position of a particle, the less accurately we know its momentum, and vice versa.

THE MEANING
OF UNCERTAINTY

These startling conclusions were published in the *Zeitschrift für Physik* in 1927, but while theorists such as Dirac and Bohr, familiar with the new equations of quantum mechanics, appreciated their significance at once, many experimenters saw Heisenberg's claim as a challenge to their skills. They imagined that he was saying that their experiments weren't good enough to measure both position and momentum at the same time, and tried to conceive experiments to prove him wrong. But this was a futile aim, since that wasn't what he had said at all.

This misconception still arises today, partly because of the way the idea of uncertainty is often taught. Heisenberg himself used the idea of observing an electron to make his point. We can only see things by looking at them, which involves bouncing photons of light off them and into our eyes. A photon doesn't disturb an object like a house very much, so we don't expect the house to be affected by looking at it. For an electron, though, things are rather different. To start with, because an electron is so small we have to use electromagnetic energy with a short wavelength in order to see it (with the aid of experimental apparatus) at all. Such gamma radiation is very energetic, and any photon of gamma radiation that bounces off an electron and can be detected by our experimental apparatus will drastically change the position and momentum of the electron—if the electron is in an atom, the very act of observing it with a gamma ray microscope may knock it out of the atom altogether.

All this is true enough, and it does give a general idea of the impossibility of measuring precisely both the position and momentum of an electron. But what the uncertainty principle tells us is that, according to the fundamental equation of quantum mechanics, there is no such thing as an electron that possesses both a precise momentum and a precise position.

This has far-reaching implications. As Heisenberg said at the end of his paper in the *Zeitschrift,* "We *cannot* know, as a matter of principle, the present in all its details." This is where quantum theory cuts free from the determinacy of classical ideas. To Newton, it would be possible to predict the entire course of the future if we knew the position and momentum of every particle in the universe; to the modern physicist, the idea of such perfect prediction is meaningless because we cannot know the position and momentum of even *one* particle precisely. The same conclusion comes out of all the different versions of the equations, the wave mechanics, the Heisenberg-Born-Jordan matrices, and Dirac's q numbers, although Dirac's approach, which carefully avoids any physical comparisons with the everyday world, seems the most appropriate. Indeed, Dirac very nearly came to the uncertainty relation before Heisenberg. In a paper for the *Proceedings of the Royal Society* in December 1926 he pointed out that in quantum theory it is impossible to answer any question that refers to numerical values of both q and p, although "one would expect, however, to be able to answer questions in which only the q or only the p are given numerical values."

It was only in the 1930s that the philosophers took up the implications of these ideas for the concept of causality—the idea that every event is caused by some other specific event—and the puzzle of predicting the future. Meanwhile, although the uncertainty relations had been derived from the fundamental equations of quantum mechanics, some influential experts began to teach quantum theory by starting out from the uncertainty relations. Wolfgang Pauli was probably the key influence in this trend. He wrote a major encyclopedia article on quantum theory that began with the uncertainty relations, and he encouraged a colleague, Herman Weyl, to begin his textbook *Theory of Groups and Quantum Mechanics* in much the same way. This book was first published in German in 1928 and in English (by Methuen) in 1931. Together, the book and Pauli's article set the tone for a generation of standard texts. Students raised on those texts became, in some cases, professors in their turn, and passed on the same style of teaching to subsequent generations. As a result students at university today are still, more often than not, introduced to quantum theory via the uncertainty relations.*

This is a peculiar accident of history. After all, the basic equations of quantum theory lead on to the uncertainty relations, but if you start with uncertainty there is no way to work out the basic quantum equations. What's worse, the only way to introduce uncertainty without the equations is to use examples like the gamma-ray microscope for observing electrons, and this immediately makes people think that uncertainty is all about experimental limitations, not a fundamental truth about the nature of the universe. You have to learn one thing, then backtrack to learn something else, then move forward to discover just what it was that you learned about first of all. Science is not always logical, nor are science teachers. The result has been generations

---

*This does make for a delightful coincidence, though. According to this way of approaching quantum theory, the most important things are the p's and q's of the uncertainty relation. Everyone knows the old expression "mind your p's and q's," which means "take care." The expression probably comes from an admonition to children learning the alphabet, or to printers' apprentices working with movable type, to watch out for the fiddly bits on the tails of these letters (*Brewer's Dictionary of Phrase and Fable,* Cassell, London, 1981), but it could now be taken as the motto of quantum theory. As far as I know, the choice of these letters in the quantum equations was, however, no more than a coincidence.

of confused students and misconceptions about the uncertainty principle—misconceptions that you don't share, because you have discovered things in the right order. However, if we are not too worried about the scientific intricacies, and we want to get our teeth into the strangeness of the quantum world, it does make a lot of sense to start out on an exploration of that world with a striking example of its peculiar nature. For the rest of this book, the uncertainty principle will be just about the *least* peculiar thing you come across.

THE COPENHAGEN INTERPRETATION

An important aspect of the uncertainty principle, which doesn't always get the attention it deserves, is that it does *not* work in the same sense forward and backward in time. Very few things in physics "care" which way time flows, and it is one of the fundamental puzzles of the universe we live in that there should indeed be a definite "arrow of time," a distinction between the past and the future. The uncertainty relations tell us that we cannot know position and momentum at the same time, and therefore we cannot predict the future—the future is inherently unpredictable and uncertain. But it is quite within the rules of quantum mechanics to set up an experiment from which it is possible to calculate backward and work out exactly what the position and momentum of an electron, say, *was* at some time in the past. The future is inherently uncertain—we do not know exactly where we are going; but the past is clearly defined—we do know exactly where we have come from. To paraphrase Heisenberg, "We *can* know, as a matter of principle, the past in all its details." This precisely fits in with our everyday experience of the nature of time, moving from a known past into an uncertain future, and it is a feature of the quantum world at its most fundamental. This may be linked to the arrow of time we perceive in the universe at large; its more bizarre possible implications will be discussed later.

While the philosophers slowly began to grapple with such intriguing implications of the uncertainty relations, to Bohr they came like a shaft of light illuminating the concepts he had been groping toward for some time. The idea of complementarity, that *both* wave and particle pictures are necessary to understand the quantum world (although in fact an electron, say, is *neither* wave nor particle), found a mathematical formulation in the uncertainty relation that said that position and momentum could not both be known precisely, but formed complementary and in a sense mutually exclusive aspects of reality. From July 1925 until September 1927 Bohr published hardly anything on quantum theory, and then he presented a lecture in Como, Italy, which introduced the idea of complementarity and what is known as the "Copenhagen interpretation" to a wide audience. He pointed out that whereas in classical physics we imagine a system of interacting particles to function, like clockwork, regardless of whether or not they are observed, in quantum physics the observer interacts with the system to such an extent that the system cannot be thought of as having independent existence. By choosing to measure position precisely, we force a particle to develop more uncertainty in its momentum, and vice versa; by choosing an experiment to measure wave properties, we eliminate particle features, and no experiment reveals both particle and wave aspects at the same time; and so on. In classical physics, we can describe the positions of particles precisely in space-time, and forecast their behavior equally precisely; in quantum physics we cannot, and in this sense even relativity is a "classical" theory.

It took a long time for these ideas to be developed and for their significance to sink in. Today, the key features of the Copenhagen interpretation can be more easily explained, and understood, in terms of what happens when a scientist makes an experimental observation. First, we have to accept that the very act of observing a thing changes it, and that we, the observers, are in a very real sense part of the experiment—there is no clockwork that ticks away regardless of whether we look at it or not. Secondly, all we know about are the results of experiments. We can look at an atom and see an electron in energy state A, then look again and see an electron in energy state B. We guess that the electron jumped from A to B, perhaps because we looked at it. In fact, we cannot even say for sure that this is the same electron, and we cannot make any statement about what it was doing when we were not looking at it. What we can learn from experiments, or from the equations of quantum theory, is the probability that if we look at a system once and get answer A then the next time we look we will get answer B. We can say nothing at all about what happens when we are not looking, and how the system gets from A to B, if indeed it does. The "damned quantum jumping" that so disturbed Schrödinger is purely our interpretation of why we get two different answers to the same experiment, and it is a false interpretation. Sometimes things are found to be in state A, sometimes in state B, and the question of what lies in between, or how they get from one state to another, is completely meaningless.

This is the really fundamental feature of the quantum world. It is interesting that there are limits to our knowledge of what an electron is doing when we *are* looking at it, but it is absolutely mind-blowing to discover that we have no idea at all what it is doing when we are not looking at it.

In the 1930s, Eddington provided what are still some of the best physical examples of what this means, in his book *The Philosophy of Physical Science*. He stressed that what we perceive, what we "learn" from experiments, is highly colored by our expectations, and he provides an example, disturbing in its simplicity, to pull the rug from under those perceptions. Suppose, he says, that an artist tells you that the shape of a human head is "hidden" in a block of marble. Absurd, you say. But then the artist, chipping away at the marble with nothing more subtle than a hammer and chisel, reveals the hidden form. Is this the way that Rutherford "discovered" the nucleus? "The discovery does not go beyond the waves which represent the knowledge we have of the nucleus," says Eddington, for nobody has ever seen an atomic nucleus. All we see are the results of experiments, which we interpret in terms of the nucleus. Nobody found a positron until Dirac suggested they might exist; today physicists claim to know of a greater number of so-called fundamental particles than there are distinct elements in the periodic table. In the 1930s, physicists were intrigued by the prediction of another new particle, the neutrino, required in order to explain subtleties of the spin interactions in some radioactive decays. "I am not much impressed by the neutrino theory," said Eddington, "I do not believe in neutrinos." But "dare I say that experimental physicists will not have sufficient ingenuity to *make* neutrinos?"

Since then, neutrinos have indeed been "discovered" in three different varieties (plus their three different anti-varieties) and other kinds are postulated. Can Eddington's doubts really be taken at face value? Is it possible that the nucleus, the positron and the neutrino did *not* exist until experimenters discovered the right sort of chisel with which to reveal their form? Such speculations strike at the roots of sanity, let alone our concept of reality. But they are quite sensible questions to ask in the quantum world. If we follow the quantum recipe book correctly, we can perform an experiment that produces a set of pointer readings that we interpret as indicating the existence of a certain kind of particle. Almost every time we follow the same recipe, we get the same set of pointer readings. But the interpretation in terms of particles is all in the mind, and may be no more than a consistent delusion. The equations tell us nothing about what the particles do when we do not look at them, and before Rutherford nobody ever looked at a nucleus, before Dirac nobody even imagined the existence of a positron. If we cannot say what a particle does when we are not looking at it, neither can we say if it exists when we are not looking at it, and it is reasonable to claim that nuclei and positrons did not exist prior to the twentieth century, because nobody before 1900 ever saw one. In the quantum world, what you see is what you get, and nothing is real; the best you can hope for is a set of delusions that agree with one another. Unfortunately, even those hopes are dashed by some of the simplest experiments. Remember the double-slit experiments that "proved" the wave nature of light? How can they be explained in terms of photons?

THE EXPERIMENT WITH TWO HOLES

One of the best, and best-known, teachers of quantum mechanics over the past twenty years has been Richard Feynman, of the California Institute of Technology. His three-volume *Feynman Lectures on Physics*, published in the early 1960s, provides a standard against which other undergraduate texts must be compared, and he has also been involved in popular lectures on the subject, such as his series on BBC television in 1965, published as *The Character of Physical Law*. Born in 1918, Feynman was at the peak of his prowess as a theoretical physicist in the 1940s, when he was involved in setting up the equations of the quantum version of electromagnetism, called quantum electrodynamics; he received the Nobel Prize for this work in 1965. Feynman's special place in the history of quantum theory is as a representative of the first generation of physicists to grow up with all of the basics of quantum mechanics established, and all the ground rules laid. Whereas Heisenberg and Dirac had to work in a changing environment, where new ideas did not always appear in their correct sequence and the logical relation of one concept to another (as in the case of spin) was not necessarily immediately obvious, to Feynman's generation, for the first time, all the pieces of the puzzle were present and the logic of their ordering could be seen, if not quite at a glance then certainly after a little thought and intellectual effort. So it is significant that whereas Pauli and his followers thought, more or less in the heat of the moment, that the uncertainty relations were the place to start discussing and teaching quantum theory, Feynman and those teachers in recent decades who look at the logic for themselves instead of reproducing the ideas of past generations have come up with a different starting point. The basic element of quantum theory, says Feynman on page 1 of the volume of his *Lectures* devoted to quantum mechanics, is the double-slit experiment. Why? Because this is "a phenomenon which is impossible, *absolutely* impossible, to explain in any classical way, and which has in it the heart of quantum mechanics. In reality, it contains the *only* mystery . . . the basic peculiarities of all quantum mechanics."

In all that has gone before in this book I have tried, like the great physicists of the first third of this century, to explain quantum ideas in terms of the everyday world. Now, starting with the central mystery, it is time to remove the blinkers of everyday experience, as far as possible, and explain the real world in terms of quantum mechanics. There are no analogies that we can carry over from our everyday experience into the world of the quantum, and the behavior

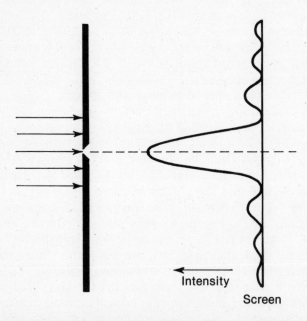

*Figure 8.1/*An electron beam passing through a single slit produces a distribution with most "particles" detected in line with the slit.

of the quantum world is not like anything familiar. Nobody knows how the quantum world behaves the way it does, all we know is that it does behave the way it does. There are just two straws to which you can cling. The first is that both "particles" (electrons) and "waves" (photons) behave in the same way—the rules of the game are consistent. The second is that, as Feynman has put it, there is only one mystery. If you can come to terms with the double-slit experiment then the battle is more than half over, since "any other situation in quantum mechanics, it turns out, can always be explained by saying, 'You remember the case of the experiment with the two holes? It's the same thing.'"*

The experiment works like this. Imagine a screen of some kind—a wall, perhaps—with two small holes in it. They can be long, narrow slits, as in Young's famous experiment with light, but small, round holes will do just as well. On one side of this wall is another wall that incorporates a detector of some kind. If we are experimenting with light, the detector might be a white surface on which we can see light and dark bands, or it might be a photographic plate that we can develop and study at leisure. If we are working with electrons, the screen might be covered in an array of electron-detectors, or we might imagine one detector on wheels that can be moved about at will to find out how many electrons are arriving at some particular spot on the screen. The details are unimportant, as long as we have some way of recording what happens at the screen. On the other side of the wall with the two holes there is a source of photons, electrons, or whatever. This might just be a lamp, or it might be an electron gun like the one that paints the picture on your TV screen; again, the details are unimportant. What happens when things go through the two holes and on to the screen—what pattern do they make at our detector?

First, step away from the quantum world of photons and electrons and look at what happens in the everyday world. It is easy to see how waves diffract through holes, by

*The Character of Physical Law, page 130.

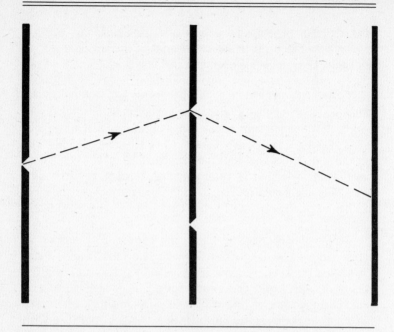

*Figure 8.2/*An electron or a photon passing through one of
a pair of slits "ought," according to common sense, to behave
in the same way as if it passed through a single slit.

using a tank of water in which all of the experiment is immersed. The source is just a device of some kind that jiggles up and down to make regular waves. The waves spread through the two holes and form a regular pattern of troughs and crests along the detector because of interference from the waves coming from each hole. If we block off one of the holes in the wall, the height of the waves on the screen varies in a simple, regular way. The biggest waves are the ones nearest the hole, across the shortest distance of the tank, and to either side the amplitude of the waves is less. The same pattern is found if we block off this hole and open up the one that was previously blocked. The intensity of the wave, which is a measure of the amount of energy that the wave carries, is proportional to the square of the height or amplitude, H^2, and shows a similar pattern for each separate hole. But when *both* holes are open, the pattern is

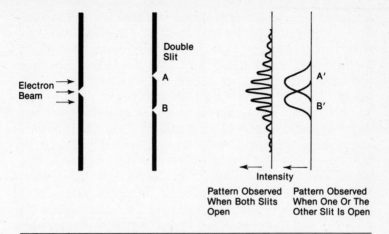

Electron Beam

Double Slit

A

B

A'

B'

Intensity

Pattern Observed When Both Slits Open

Pattern Observed When One Or The Other Slit Is Open

*Figure 8.3/*For electrons and for photons, however, experiments show that the pattern observed with both slits "open" is not the same as the pattern obtained by adding together what we see with each slit separately.

much more complex. There is indeed a large peak intensity smack in line with the two holes, but there is a very low intensity just either side of the peak, where the two sets of waves cancel out, and a pattern of highs and lows repeating alternately as we move along the screen. Mathematically, instead of finding that the intensity of both holes together is the sum of their two separate intensities (the sum of the squares), it turns out to be the square of the sum of the two amplitudes. For waves whose amplitudes are represented by H and J, say, the intensity I is *not* $H^2 + J^2$, but is given by the expression

$$I = (H + J)^2$$

which works out as

$$I = H^2 + J^2 + 2HJ$$

The extra term is the contribution due to interference from the two waves, and, making allowance for the fact that the H's and J's can be negative or positive, it precisely explains the peaks and troughs in the interference pattern.

If we carried out the same sort of experiment using large particles in the everyday world (Feynman whimsically imagines an experiment involving a machine gun shooting bullets through the holes in the wall, and buckets of sand in which to collect them spaced along the detector), we would not find any "interference term." We *would* find, after we had fired a large number of bullets through the holes, different numbers of bullets in different buckets. With only one hole open, the pattern of bullets scattered around the "screen" would be very like the pattern of the intensity variations for water waves with one hole open. But with *both* holes open, the pattern of bullets found in different bins would indeed be just the sum of the two effects from the two separate holes—most bullets in the region just behind the two holes, and a nice smooth tail-off on either side, with no peaks and troughs caused by interference. In this case, regarding each bullet as representing a unit of energy, the intensity distribution *is* given by

$$I = I_1 + I_2,$$

where I_1 corresponds to H^2 and I_2 to J^2 in the wave example. There is no interference term.

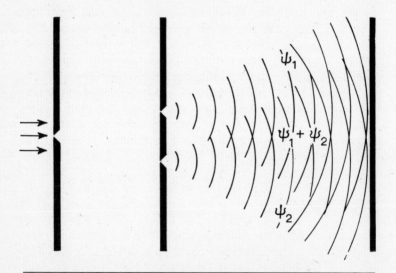

Figure 8.4/"Probability waves" seem to decide where each "particle" in the beam goes, and probability waves interfere just as water waves do (see Figure 1.3).

You know what's coming next. Now imagine the same experiments done with light and with electrons. The double-slit experiment has, of course, really been carried out in precisely this way many, many times with light, and it produces diffraction patterns just as in the wave example. The electron experiment hasn't quite been carried out in this way—there are problems making things on a small enough scale—but equivalent experiments have been carried out by scattering beams of electrons from atoms in crystals. To keep the story simple, though, I'll stick with the imaginary double-slit experiment, translating into that language the unambiguous results that come out of the real electron experiments. Just like light, the electrons too show the diffraction pattern.

So what? Isn't this just the particle/wave duality that we have learned to live with? The point is that we learned to live with it for the purposes of the quantum cookbook, but

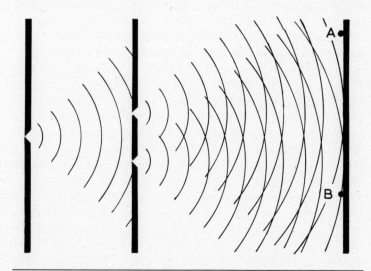

*Figure 8.5/*The rules of wave behavior are needed to assign probabilities to the appearance of an electron at A or B; yet when we look at A or B we either see an electron—a particle—or not. We don't see a wave. We cannot say what the electron is "really" doing during its passage through the apparatus.

that we did not look deeply into the implications. The time has come to do so. Schrödinger's function ψ, the variable in his wave equation, has something to do with an electron (or whatever particle the equation is describing). If ψ is a wave, it is no surprise to find that it diffracts and produces an interference pattern, and it is a simple step to show that ψ acts like the amplitude of the wave and ψ^2 acts like intensity. The diffraction pattern of the electron two-hole experiment is a pattern of ψ^2. If there are many electrons in the beam, this has a simple interpretation—ψ^2 represents the *probability* of finding an electron in some particular place. Thousands of electrons rush through the two holes, and where they end up can be predicted on a statistical basis using this interpretation of the ψ wave—Born's great contribution to quantum cookery. But what happens to each individual electron?

We can understand easily enough that a wave—a water wave, perhaps—can pass through both holes in the screen. A wave is a spread-out thing. But an electron still seems to be a particle, even if it has associated wavelike properties. It is natural to believe that each individual electron *must*, surely, go through one hole or the other. We can try experimentally the equivalent of blocking off each hole in turn. When we do, we get the usual pattern on our screen for single-hole experiments. When we open both holes together, however, we do not get the pattern produced by adding up those two patterns, as we would for bullets. Instead, we get the pattern for interference by waves. And we *still* get this pattern if we slow down our electron gun so much that only one electron at a time goes through the whole setup. One electron goes through only one hole, we would guess, and arrives at our detector; then another electron is let through, and so on. If we wait patiently for enough electrons to pass through, the pattern that builds up on our detector screen is the diffraction pattern for waves. Indeed, for electrons or photons, if we took a thousand identical experiments in a thousand different laboratories, and let one particle pass through each experiment, we could add up the thousand different results and still get an overall distribution pattern in line with diffraction, just as if we let a thousand electrons through one of those experiments together. A single electron, or a single photon, on its way through one hole in the wall, obeys the statistical laws which are only appropriate if it "knows" whether or not the other hole is open. This is the central mystery of the quantum world.

We can try cheating—shutting or opening one of the holes quickly while the electron is in transit through the apparatus. It doesn't work—the pattern on the screen is always the "right" one for the state of the holes at the instant the electron was passing through. We can try peeking, to "see" which hole the electron goes through. When the equivalent of this experiment is carried out, the result is even more bizarre. Imagine an arrangement that records which hole an electron goes through but lets it pass on its way to the detector screen. Now the electrons behave like normal, self-respecting everyday particles. We always see an electron at one hole or the other, never both at once. And now the pattern that builds up on the detector screen is exactly equivalent to the pattern for bullets, with no trace of interference. The electrons not only know whether or not both holes are open, they know whether or not we are watching them, and they adjust their behavior accordingly. There is no clearer example of the interaction of the observer with the experiment. When we try to look at the spread-out electron wave, it collapses into a definite particle, but when we are not looking it keeps its options open. In terms of Born's probabilities, the electron is being forced by our measurement to choose one course of action out of an array of possibilities. There is a certain probability that it could go through one hole, and an equivalent probability that it may go through the other; probability interference produces the diffraction pattern at our detector. When we detect the electron, though, it can only be in one place, and that changes the probability pattern for its future behavior—for that electron, it is now certain which hole it went through. But unless someone looks, nature herself does not know which hole the electron is going through.

COLLAPSING WAVES

What we see is what we get. An experimental observation is
only valid in the context of the experiment and can't be
used to fill in details of things we do not observe. You might
say that the double-slit experiment tells us that we are dealing with waves; equally, by looking only at the pattern on
the detector screen you can deduce that the apparatus has
two holes in it, not one. The whole thing is what matters—
the apparatus, the electrons, and the observer are all part of
the experiment. We cannot say that an electron goes
through either hole, without looking at the holes as it
passes (and that is a different experiment). An electron
leaves the gun and arrives at the detector, and it seems to
possess information about the whole experimental setup,
including the observer. As Feynman explained to his BBC
TV audience in 1965, if you have an apparatus that is capable of telling which hole the electron goes through, then
you can say that it either goes through one hole or the
other. But when you have no apparatus to determine
through which hole the thing goes, then you cannot say
that it goes through either one hole or the other. "To conclude that it goes either through one hole or the other when
you are not looking is to produce an error," he states. The
term "holistic" has become such a misused buzzword that I
hesitate to introduce it. However, there is no word more apt
to describe the quantum world. It *is* holistic; the parts are
in some sense in touch with the whole. And this doesn't
just mean the whole of an experimental setup. The world
seems to keep all its options, all its probabilities, open for as
long as possible. The strangest thing about the standard
Copenhagen interpretation of the quantum world is that it
is the act of observing a system that forces it to select one of
its options, which then becomes real.

In the simplest experiment with two holes, the interference of probabilities can be interpreted as if the electron
that leaves the gun vanishes once it is out of sight, and is
replaced by an array of ghost electrons that each follows a

different path to the detector screen. The ghosts interfere
with one another, and when we look at the way electrons
are detected by the screen we then find the traces of this
interference, even if we deal only with one "real" electron at
a time. However, this array of ghost electrons only describes what happens when we are not looking; when we
look, all of the ghosts except one vanish, and one of the
ghosts solidifies as a real electron. In terms of Schrödinger's
wave equation, each of the "ghosts" corresponds to a wave,
or rather a packet of waves, the waves that Born interpreted
as a measure of probability. The observation that crystalizes
one ghost out of the array of potential electrons is equivalent, in terms of wave mechanics, to the disappearance of
all of the array of probability waves except for one packet of
waves that describes one real electron. This is called the
"collapse of the wave function," and, bizarre though it is, it
is at the heart of the Copenhagen interpretation, which is
itself the foundation of quantum cookery. It is doubtful,
however, that many of the physicists, electronics engineers,
and others who happily use the recipes in the quantum
cookbook appreciate that the rules that prove so reliable in
the design of lasers and computers, or studies of genetic
material, depend explicitly on the assumption that myriad
ghost particles interfere with each other all the time, and
only coalesce into a single real particle as the wave function
collapses during an observation. What's worse, as soon as
we *stop* looking at the electron, or whatever we are looking
at, it immediately splits up into a new array of ghost particles, each pursuing their own path of probabilities through
the quantum world. Nothing is real unless we look at it, and
it ceases to be real as soon as we stop looking.

Perhaps the people who use the quantum cookbook so
happily are comforted by the familiarity of the mathematical equations. Feynman explains the basic recipe simply. In quantum mechanics, an "event" is a set of initial and
final conditions, no more and no less. An electron leaves the
gun on one side of our apparatus, and the electron arrives
at a particular detector on the other side of the holes. That
is an event. The probability of an event is given by the
square of a number which is, essentially, Schrödinger's

wave function, ψ. If there is more than one way in which the event can occur (both holes are open inside the experiment), then the probability of each possible event (the probability of the electron arriving at each chosen detector) is given by the square of the sum of the ψ's, and there is interference. But when we make an observation to find out which of the alternative possibilities actually happens (look to see which hole the electron goes through) the probability distribution is just the sum of the squares of the ψ's, and the interference term disappears—the wave function collapses.

The physics is impossible, but the math is clean and simple, familiar equations to any physicist. As long as you avoid asking what it means, there are no problems. Ask why the world should be like this, however, and even Feynman has to reply, "We have no idea." Persist in asking for a physical picture of what is going on, and you find all physical pictures dissolving into a world of ghosts, where particles only seem to be real when we are looking at them, and where even a property such as momentum or position is only an artifact of the observation. It is scarcely any wonder that many respected physicists, including Einstein, spent decades trying to find ways around this interpretation of quantum mechanics. Their attempts, which will be described briefly in the next chapter, have all failed, and each new failure of attempts to disprove the Copenhagen interpretation has strengthened the basis for this picture of a ghostly world of probabilities, paving the way to move beyond quantum mechanics and to develop a new picture of the holistic universe. The basis for that new picture is the ultimate expression of the concept of complementarity, but there is one final bullet to bite on before we can look at the implications.

COMPLEMENTARITY RULES

General relativity and quantum mechanics are usually represented as the twin triumphs of twentieth-century theoretical science, and the Holy Grail of physicists today is a true unification of the two into one grand theory. Their efforts, as we shall see, are certainly providing deep insights into the nature of the universe. But those efforts do not seem to take account of the fact that in a strict sense the two pictures of the world may be irreconcilable.

In his very first exposition of what became known as the Copenhagen interpretation, back in 1927, Bohr stressed the contrast between descriptions of the world in terms of a pure space-time coordination and absolute causality, and the quantum picture, where the observer interferes with and is a part of the system that is being observed. Coordinates in space-time represent position; causality depends on knowing precisely where things are going, essentially on knowing their momentum. Classical theories assume that you can know both at once; quantum mechanics shows us that precision in space-time coordination has to be paid for in terms of uncertainty of momentum, and therefore of causality. General relativity is a classical theory, in this sense, and cannot be regarded as the equal of quantum mechanics as a fundamental description of the universe. If and when we find a conflict between the two, it is to quantum theory that we must turn for the best description of the world in which we live.

But what is the world in which we live? Bohr suggested that the very idea of a unique "world" may be misleading, and offered another interpretation of the experiment with two holes. Even in that simple experiment, of course, there are many paths that an electron or photon can choose through *each* of the two holes. But for simplicity, let's pretend there are just two possibilities, that the particle goes through hole A or through hole B. Bohr suggested that we might think of each possibility as representing a different world. In one world, the particle goes through hole A; in the other, it goes through hole B. The real world, the world that we experience, is neither of these simple worlds, however. Our world is a hybrid combination of the two possible worlds corresponding to the two routes for the particle, and each world interferes with the other. When we look to see which hole the particle goes through, there is now only one world because we have eliminated the other possibility, and in that case there is no interference. It isn't just ghost electrons that Bohr conjures out of the quantum equations, but ghost realities, ghost *worlds* that only exist when we are not looking at them. Imagine this simple example elaborated to cover not just two worlds united by a two-hole experiment, but a myriad array of ghost realities corresponding to all the myriad ways every quantum system in the entire universe could "choose" to jump: every possible wave function for every possible particle; every allowed value of Dirac's q numbers. Combine that with the puzzle that an electron at hole A *knows* whether hole B is open or closed, and that in principle it knows the quantum state of the entire universe, and it is easy to see why the Copenhagen interpretation was attacked with such vigor by some of the experts who understood its deepest implications, even while other experts, though disturbed by the implications, found the interpretation compelling, and while lesser mortals, not bothered by the deep implications, happily proceeded to use the quantum cookbook, collapsing wave functions and all, to transform the world we live in.

CHAPTER NINE

PARADOXES
AND POSSIBILITIES

Each attack on the Copenhagen interpretation has strengthened its position. When thinkers of the caliber of Einstein try to find flaws with a theory, but the defenders of the theory are able to refute all of the attackers' arguments, that theory must emerge the stronger for its trial. The Copenhagen interpretation is definitely "right" in the sense that it works; any better interpretation of the quantum rules must include the Copenhagen interpretation as a working view that enables experimenters to predict the outcome of their experiments—at least in a statistical sense—and enables engineers to design working laser systems, computers, and so on. There is no point in going over all of the groundwork that resulted in the refutation of all of the counterproposals to the Copenhagen interpretation, a job that has been well done by others. Perhaps, though, the most important point to notice is one made by Heisenberg in his book *Physics and Philosophy*, back in 1958. All of the counterproposals, Heisenberg stressed, are "compelled to sacrifice the essential symmetry of quantum theory (for instance, the symmetry between waves and particles or between position and velocity). Therefore, we may well suppose that the Copenhagen interpretation cannot be avoided if these symmetry properties . . . are held to be a genuine feature of nature; and every experiment yet performed supports this view" (page 128).

There is an *improvement* on the Copenhagen interpretation (*not* a refutation or counterproposal) that still includes this essential symmetry, and that best-buy picture of quantum reality will be described in Chapter Eleven. However, it is hardly surprising that Heisenberg failed to mention it in a book published in 1958, since the new picture was only just being developed at that time, by a PhD student in the United States. Before we get on to that, however, it is right to trace the path of a combination of theory and experiment that by 1982 had established beyond any doubt the accuracy of the Copenhagen interpretation as a working view of quantum reality. The story starts with Einstein, and ends in a physics laboratory in Paris more than fifty years later; it is one of the great stories of science.

THE CLOCK IN THE BOX

The great debate between Bohr and Einstein on the interpretation of quantum theory began in 1927 at the fifth Solvay Congress, and continued until Einstein's death in 1955. Einstein also corresponded with Born on the subject, and a flavor of the debate can be gleaned from *The Born-Einstein Letters*. The debate centered on a series of imaginary tests of the predictions of the Copenhagen interpretation—not real experiments carried out in the lab, but "thought experiments." The game was that Einstein would try to think up an experiment in which it would be theoretically possible to measure two complementary things at once, the position and mass of a particle, or its precise energy at a precise time, and so on. Bohr, and Born, would then try to show how Einstein's thought experiment simply could not be carried out in the way required to pull the rug from under the theory. One example, the "clock in the box" experiment, will serve to show how the game was played.

Imagine a box, said Einstein, that has a hole in one wall covered by a shutter that can be opened and closed again under the control of a clock inside the box. Apart from the clock and shutter mechanism, the box is filled with radiation. Set up the apparatus so that at some precise, predetermined time on the clock the shutter will open and allow one photon to escape before it closes again. Now weigh the box, wait for the photon to escape, and weigh the box again. Because mass is energy, the difference in the two weights

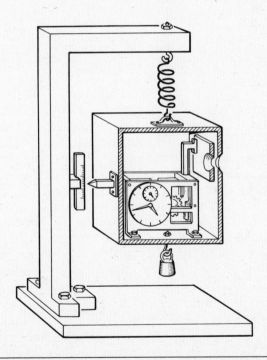

*Figure 9.1/*The "clock in the box" experiment. The paraphernalia needed to make the experiment practicable (weights, springs, and so on) always make it impossible to remove the uncertainty from the measurement of energy and time together (see text).

tells us the energy of the photon that escaped. So we know, in principle, the exact energy of the photon and the exact time that it passed through the hole, refuting the uncertainty principle.

Bohr, as always in these arguments, won the day by looking at the practical details of how the measurements could be carried out. The box must be weighed, so it must be suspended by a spring, for example, in a gravitational field. Before the photon escapes from the box, the mythical experimenter notes the position of a pointer, firmly fixed to the box, against a scale. After the photon escapes, the experimenter can, in principle, add weights to the box to restore the pointer to the same place. But this in itself involves the uncertainty relations. The position of the pointer can only be determined to within the limits set by Heisenberg's relation, and there is an uncertainty in the momentum of the box associated with that uncertainty in the position of the pointer. The greater the accuracy of the measurement of the weight of the box, the greater is the uncertainty in the all-important knowledge of its momentum. Even if you try to restore the original situation by adding a small weight to the box to stretch the spring back to its original position, and measure the extra weight to determine the energy of the escaping photon, you can never do better than reducing the uncertainty to the limits allowed by the Heisenberg relation, in this case $\Delta E \, \Delta t > \hbar$.

The details of this and the other thought experiments involved in the Einstein-Bohr debate can be found in Abraham Pais's *Subtle Is the Lord . . . .* Pais stresses that there is nothing fanciful in Bohr's insistence on a full and detailed description of the mythical experiments—in this case, heavy bolts that fix the framework of the balance into place, the spring that allows mass to be measured but must thereby allow the box to move, the little weight that has to be added, and so on. The results of all experiments have to be interpreted in terms of classical language, the language of everyday reality. We *could* fix the box rigidly into place so that we had no uncertainty about its position, but then it would be impossible to measure the change in mass. The dilemma of quantum uncertainty arises because we try to express quantum ideas in everyday language, and that is why Bohr stressed the nuts and bolts of the experiments.

THE "EPR PARADOX"

Einstein accepted Bohr's criticisms of this and other thought experiments, and by the early 1930s he had turned to a new kind of imaginary test of the quantum rules. The basic idea behind this new approach was to use experimental information about one particle to deduce the properties, such as position and momentum, of a second particle. This version of the debate was never resolved in Einstein's lifetime, but it has now been successfully tested, not by an improved thought experiment but by a real experiment in the lab. Once again, Bohr wins and Einstein loses.

During the early 1930s, Einstein's personal life was in a turmoil. He had to leave Germany because of the threat of persecution by the Nazi regime. By 1935 he was settled in Princeton, and in December 1936 his second wife, Elsa, died after a long illness. Amid all this turmoil, he continued to puzzle over the interpretation of quantum theory, defeated by Bohr's arguments but not convinced in his heart that the Copenhagen interpretation, with its inherent uncertainty and lack of strict causality, could be the last word as a valid description of the real world. Max Jammer has described in exhaustive detail the various twists and turns of Einstein's mind on this subject at that time, in *The Philosophy of Quantum Mechanics*. Several threads came together in 1934 and 1935, when Einstein worked in Princeton with Boris Podolsky and Nathan Rosen on a paper presenting what has become known as the "EPR Paradox," even though it does not really describe a paradox at all.*

*A. Einstein, B. Podolsky, and N. Rosen, "Can quantum-mechanical description of physical reality be considered complete?" *Physical Review,* volume 47, pp. 777–780, 1935. The paper is among those reprinted in the volume *Physical Reality,* edited by S. Toulmin, Harper & Row, 1970.

The point of the argument was that, according to Einstein and his collaborators, the Copenhagen interpretation had to be *incomplete*—that there really is some underlying clockwork that keeps the universe running, and that only gives the *appearance* of uncertainty and unpredictability at the quantum level, through statistical variations.

Imagine two particles, said Einstein, Podolsky, and Rosen, that interact with one another and then fly apart, not interacting with anything else at all until the experimenter decides to investigate one of them. Each particle has its own momentum, each is located at some position in space. Even within the rules of quantum theory, we are allowed to measure *precisely* the total momentum of the two particles, added together, and the distance between them, at the time they were close together. When, much later, we decide to measure the momentum of one of the particles we know, automatically, what the momentum of the other one must be, because the total must be unchanged. Alternatively, we could have measured the precise position of the first particle and, in the same way, deduced the position of the second particle. Now, it is one thing to argue that the physical measurement of the momentum of particle A destroys knowledge of its *own* position, so that we cannot know its precise position, and that similarly the physical measurement of position of particle A disturbs its momentum, which remains unknown. But it seemed quite another thing entirely, to Einstein and his colleagues, to argue that the state of particle B depended on which of the two measurements we choose to make on particle A. How can particle B "know" whether it should have a precisely defined momentum *or* a precisely defined position? It seemed as if, in the quantum world, the measurements we make on a particle *here* affect its partner *there,* in violation of causality, an instantaneous "communication" traveling across space, something called "action at a distance."

If you accepted the Copenhagen interpretation, the EPR paper concluded, it "makes the reality of [position and momentum in the second system] depend upon the process of measurement carried out on the first system which does not disturb the second system in any way. *No reasonable* *definition of reality could be expected to permit this.*"* This is where the team diverged from most of their colleagues, and from all of the Copenhagen school. Nobody disagreed with the logic of the argument, but they did disagree on what constitutes a "reasonable" definition of reality. Bohr and his colleagues could live with a reality in which the position and momentum of the second particle had no objective meaning until they were measured, regardless of what you did to the first particle. A choice had to be made between a world of objective reality and the quantum world, of that there was no doubt. But Einstein remained in a very small minority in deciding that of the two options open he would cling to objective reality and reject the Copenhagen interpretation.

But Einstein was an honest man, always ready to accept sound experimental evidence. If he had lived to see it, he would certainly have been persuaded by the recent experimental tests of what is effectively a kind of EPR effect that he was wrong. Objective reality does *not* have any place in our fundamental description of the universe, but action at a distance, or acausality, does have such a place. The experimental verification of this is so important that it deserves a chapter to itself. But first, for completeness, we ought to look at some of the other paradoxical possibilities inherent in the quantum rules—particles that travel backward in time and, at last, Schrödinger's famous half-dead cat.

TIME TRAVEL

Physicists often use a simple device to represent the movement of particles through space and time on a piece of paper or on the blackboard. The idea is simply to represent the flow of time by the direction up the page, from bottom to top, and motion in space across the page. This squeezes

---

*Quoted by Pais, page 456.

three space dimensions into one, but produces patterns that are immediately familiar to anyone who has dealt with graphs, with time corresponding to the "y" axis and space to the "x" axis. These space-time diagrams first appeared as an invaluable tool of modern physics in relativity theory, where they can be used to represent many of the peculiarities of Einstein's equations in geometrical terms that are sometimes easier to manipulate and often easier to understand. They were taken over into particle physics by Richard Feynman in the 1940s, and in that context they are usually called "Feynman diagrams"; in the quantum world of particles, the space and time representation can also be replaced by a description in term of momentum and energy, which is more relevant when dealing with collisions between particles, but I'll stick with a simple space-time description here.

The track of an electron is represented on a Feynman diagram by a line. An electron that sits in one place and never moves produces a line that moves straight up the page, corresponding to motion in the time direction only;

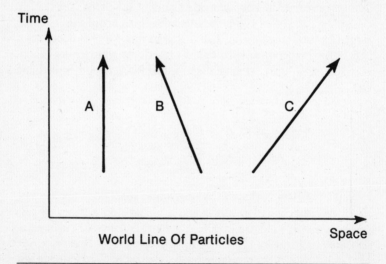

Figure 9.2/The motion of a particle through time and space can be represented as a "world line."

an electron that slowly changes its position, as well as being carried along by the flow of time, is represented by a line at a slight angle to the line straight up the page, and a fast-moving electron makes a bigger angle with the "world line" of a stationary particle. The motion in space can be in either direction, to left or right, and the line may zigzag if the electron is deflected by collisions with other particles. But in the everyday world, or the world of simple space-time diagrams in relativity theory, we would not expect the world line to turn back and progress down the page, because that would correspond to movement backward in time.

Sticking with electrons as our example, we can plot a simple Feynman diagram showing how an electron moves through space and time, collides with a photon and changes its direction, then emits a photon and recoils in yet

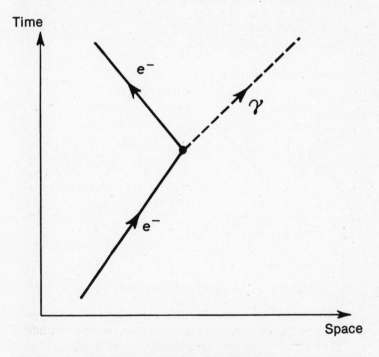

Figure 9.3/ An electron moves through space and time, emits a photon (γ-ray) and recoils at an angle.

another direction. Photons are crucially important in this
description of particle behavior, because they act as the carriers of the electric force. When two electrons come near

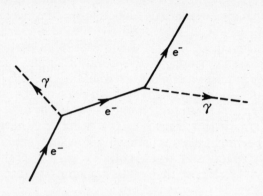

Figure 9.4/Part of the life history of an
electron involving two photon interactions.

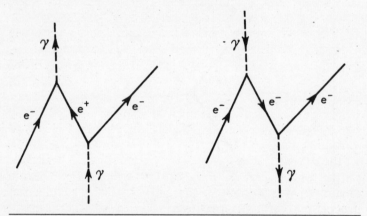

Figure 9.5/On the left, a gamma ray produces an
electron/positron pair, and the positron later meets
another electron and annihilates with it to make
another photon. On the right, a single electron moves
in a zigzag through space-time and interacts with
two photons, just as in Figure 9.4. But for part of its
life, this electron moves backward in time. The
two pictures are mathematically equivalent.

one another, they repel each other and move apart again because of the electric force between their like charges. The Feynman diagram for such an encounter sees two electron world lines converging, then a photon leaving one electron (which recoils away) and being absorbed by the other electron (which is pushed in the other direction).* Photons are the carriers of the electric field. But they can do more than this. Dirac showed that a sufficiently energetic photon could produce an electron and a positron out of the vacuum, converting its energy into their mass. The positron (negative-energy electron "hole") will be short lived, because it is bound to meet another electron soon and the pair will annihilate in a burst of energetic radiation, which, for simplicity, we can represent as a single photon.

Again, the whole interaction can be represented simply in a Feynman diagram. A photon traveling through space and time spontaneously creates an electron/positron pair; the electron proceeds on its way; the positron meets another electron and disappears; another photon leaves the scene. But the dramatic discovery that Feynman made in 1949 is that the space-time description of a positron moving forward in time is *exactly* equivalent to the same mathematical description of an electron moving backward through time along the same track in the Feynman diagram. In addition, because photons are their own antiparticles, there is no difference in this description between a photon moving forward in time and one moving backward in time. For all practical purposes, we can take the arrows off the photon tracks in the diagram, and reverse the arrow on the positron track to make it an electron. The same Feynman diagram now tells us a different story. An electron proceeding through space and time meets an energetic photon, absorbs it, and is scattered *backward in time* until it emits an energetic photon and recoils in such a way that it moves forward in time again. Instead of three particles,

---

*This is, of course, a great simplification. We should imagine the electron pair actually exchanging many photons as they interact. In the same way, in what follows I refer to "a photon" creating a positron/electron pair, where in reality we would be dealing with more than one photon, perhaps a pair of colliding gamma rays or an even more complex situation.

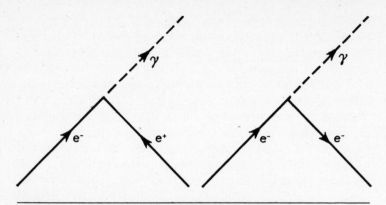

*Figure 9.6/*In general, the annihilation of a particle/
antiparticle pair can also be described as a scattering
event so severe that it sends the particle backward in time.

two electrons, and a positron in a complicated dance, we
have one particle, one electron that zigzags its way through
space *and* time, colliding with photons here and there
along the way.

In terms of the geometry of the diagrams, there is a
clear similarity between the example of an electron that absorbs a low-energy photon and changes its path slightly,
then emits the photon and changes direction again, and the
electron that is so violently scattered by the photon interaction that it travels backward in time for part of its life. In
both cases, there is a zigzag line with three straight sections and two corners. The difference is just that in the
second case the corners are much sharper than in the first.
It was John Wheeler who first had the intuition that both
zigzag patterns represented the same kind of event, but it
was Feynman who proved the exact mathematical identity
between the two cases.

There is a great deal to absorb here, more even than
meets the eye at first glance. So let's take it slowly, piece by
piece.

First, I threw in that remark about the photon being its
own antiparticle, so that we can remove the arrows from
the photon tracks. A photon going forward in time is the

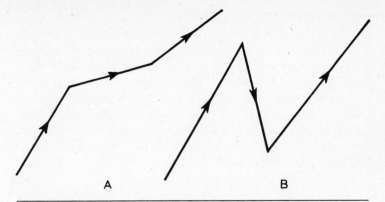

Figure 9.7/ Richard Feynman established the mathematical equivalence of all space-time diagrams with the double bend.

same as an antiphoton going backward in time, but an anti-photon is a photon, so a photon going forward in time is the same as a photon going backward in time. Did that strike you as odd? It should have. Apart from anything else, it means that when we see an atom in an excited state emit energy and fall into the ground state we might just as well say that the electromagnetic energy traveling backward in time arrived at the atom and caused the transition. This is a little tricky to imagine, because now we are not talking about an individual photon moving in a straight line through space, but an expanding spherical shell of electromagnetic energy, a wavefront spreading out in all directions from the atom and getting distorted and scattered as it goes. Reversing that picture produces a universe in which a perfectly spherical wavefront centered on our chosen atom has to be created by the universe, out of a series of scattering processes working together and focused down to converge on that one particular atom.

I don't want to go too deeply into this line of thought, because it takes us away from quantum theory and into cosmology. But it does have deep implications for our understanding of time and why we see time flowing only in one direction. Very simply, the radiation emitted by an atom now is going to be absorbed by other atoms later on. This is only possible because most of those other atoms are in their ground state, which means that the future of the universe is cold. The asymmetry, which we see as the arrow of time, is the asymmetry between the colder and hotter epochs of the universe. It is easier to arrange for a cold future to do the necessary absorbing if the universe is expanding, because the expansion itself produces a cooling effect, and we do live in an expanding universe. The nature of time as we see it may, therefore, be intimately linked with the nature of the expanding universe.*

EINSTEIN'S TIME

But what does the photon itself "see" as the arrow of time? We learn from relativity theory that moving clocks run slow, and that they run slower the closer they get to the speed of light. Indeed, *at* the speed of light, time stands still, and the clock stops. A photon, naturally, travels at the speed of light, and this means that for a photon time has no meaning. A photon that leaves a distant star and arrives at the earth may spend thousands of years on the journey, measured by clocks on earth, but takes no time at all as far as the photon is concerned. A photon of the cosmic background radiation has, from our point of view, been traveling through space for perhaps 15 thousand million years since the Big Bang in which the universe as we know it began, but to the photon itself the Big Bang and our present are the same time. The photon's track on a Feynman diagram has no arrow on it not only because the photon is its own antiparticle, but because motion through time has no meaning for the photon—and *that* is why it is its own antiparticle.

---

*These ideas are discussed in more detail, but in clear, nonmathematical language, in Chapter 6 of Jayant Narlikar's *The Structure of the Universe*, Oxford University Press, 1977. Paul Davies's *Space and Time in the Modern Universe* (Cambridge University Press, 1977) goes into even more detail, and some of the math can be found in *The Ultimate Fate of the Universe*, by J. N. Islam (Cambridge University Press, 1983).

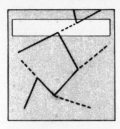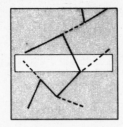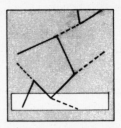

*Figure 9.8/*If all the particle tracks were somehow fixed in space-time, we might see an illusion of movement and interactions as our perception shifted from now (right-hand picture) forward through time and up the page. Is the dance of the particles just an illusion caused by our perception of the flow of time?

The mystics and popularizers who seek to equate Eastern philosophy with modern physics seem to have missed this point, which tells us that everything in the universe, past, present, and future, is connected to everything else, by a web of electromagnetic radiation that "sees" everything at once. Of course, photons can be created and destroyed, so the web is incomplete. But the reality is a photon track in space-time, linking my eye with, perhaps, the Pole Star. There is no real movement of time that sees a track developing from the star to my eye; that is just my perception from my viewpoint. Another, equally valid viewpoint sees that track as an eternal feature around which the universe changes, and during those changes in the universe one of the things that happens is that my eye and the Pole Star happen to be at opposite ends of the track.

What about the other particle tracks in the Feynman diagrams? How "real" are they? We can say much the same thing about them. Imagine a Feynman diagram that encompassed all of space and time, with the track of every particle laid out on it. Now imagine viewing that diagram through a narrow slot that only allows a limited segment of time to be scanned, and move the slot steadily up the page. Through the slot, we see a complex dance of interacting particles, pair production, annihilation, and far more complex events, an ever-changing panorama. All we are doing, though, is scanning something that is fixed in space and time. It is our perception that alters, not the underlying reality. Because we are locked into a steadily moving viewing slot, we see a positron moving forward in time rather than an electron moving backward in time, but both interpretations are equally real. John Wheeler has gone further, pointing out that we could imagine *all* of the electrons in the universe being connected by interactions to form a highly complex zigzag path through space-time, forward and backward. This was part of the original flash of inspiration that led to Feynman's definitive work—the image of "a single electron shuttling back and forth, back and forth, back and forth on the loom of time to weave a rich tapestry containing perhaps all the electrons and positrons in the world."* In such a picture, every electron everywhere in the universe would be simply a different segment of just one world line, the world line of the only "real" electron.

That idea won't work in our universe. To make it work you would expect to find as many reversed segments of the world line, as many positrons as there are forward segments—electrons. The idea of a fixed reality with our view being the only thing that changes also probably doesn't work at this simple level—how could it be reconciled with the uncertainty principle?† But together these ideas represent a far better grasp of the nature of time than our everyday experience provides. The flow of time in the everyday world is a statistical effect, largely caused by the expansion of the universe from a hotter to a cooler state. But even at that level the equations of relativity permit time travel, and the concept can be very easily understood in terms of space-time diagrams.‡

---

*Quote, based on Wheeler's explanation of his vision, from Banesh Hoffmann's *The Strange Story of the Quantum*. Pelican edition, 1963, page 217.

†Feynman really went much further than I have indicated in this simple exposition and did develop a treatment of world lines including probabilities, thereby producing a new version of quantum mechanics that was soon shown, by Freeman Dyson, to be exactly equivalent to the original versions of the theory in its results, but has since proved a much more powerful mathematical tool. More of this later.

‡The implications of relativity theory for our understanding of the universe.

Motion in space can proceed in any direction and back again. Motion in time only proceeds in one direction in the everyday world, whatever seems to be going on at the particle level. It's hard to visualize the four dimensions of space-time, each at right angles to the other, but we can leave out one dimension and imagine what this strict rule would mean if it applied to one of the three dimensions we are used to. It's as if we were allowed to move either up or down, either forward or back, but that sideways motion was restricted to shuffling to the left, say. Movement to the right is forbidden. If we made this the central rule in a children's game, and then told a child to find a way of reaching a prize off to the right-hand side ("backward in time") it wouldn't take too long for the child to find a way out of the trap. Simply turn around to face the other way, swapping left for right, and then reach the prize by moving to the left. Alternatively, lie down on the floor so that the prize is in the "up" direction with reference to your head. Now you can move both "up" to grasp the prize and "down" to your original position, before standing up again and returning your personal space orientation to that of the bystanders.* The technique for time travel allowed by relativity theory is very similar. It involves distorting the fabric of space-time so that in a local region of space-time the time axis points in a direction equivalent to one of the three space directions in the undistorted region of space-time. One of the other space directions takes on the role of time, and by swapping space for time such a device would make true time travel, there and back again, possible.

American mathematician Frank Tipler has made the calculations that prove such a trick is theoretically possible. Space-time can be distorted by strong gravitational fields,

---

and the implications for time travel, are covered in more detail in my book *Spacewarps* (Delacorte, New York; and Pelican, London, 1983).

*I tried this on a few children and adults, separately. About half the children spotted the trick, but very few of the adults. Those who *didn't* spot it complained of cheating; the fact is, according to Einstein's equations, nature herself is not above this kind of cheating.

and Tipler's imaginary time machine is a very massive cylinder, containing as much matter as our sun packed into a volume 100 km long and 10 km in radius, as dense as the nucleus of an atom, rotating twice every millisecond and dragging the fabric of space-time around with it. The surface of the cylinder would be moving at half the speed of light. This isn't the sort of thing even the maddest of mad inventors is likely to build in his backyard, but the point is that it is allowed by all the laws of physics that we know. There is even an object in the universe that has the mass of our sun, the density of an atomic nucleus, and spins once every 1.5 milliseconds, only three times slower than Tipler's time machine. This is the so-called "millisecond pulsar," discovered in 1982. It is highly unlikely that this object is cylindrical—such extreme rotation has surely flattened it into a pancake shape. Even so, there must be some very peculiar distortions of space-time in its vicinity. "Real" time travel may not be impossible, just extremely difficult and very, very unlikely. That thin end of what might be a very large wedge may, however, make the normality of time travel at the quantum level seem a little more acceptable. *Both* quantum theory and relativity theory permit time travel, of one kind or another. And anything that is acceptable to both those theories, no matter how paradoxical that something may seem, has to be taken seriously. Time travel, indeed, is an integral part of some of the stranger features of the particle world, where you can even get something for nothing, if you are quick about it.

SOMETHING FOR NOTHING

In 1935 Hideki Yukawa, then a twenty-eight-year-old physics lecturer at Osaka University, suggested an explanation for the way the neutrons and protons in an atomic nucleus could be held together in spite of the positive charge that would tend to blow the nucleus apart through the electric force. Clearly, there must be another, stronger force that overcomes the electric force under the right circumstances. The electric force is carried by the photon, so this strong nuclear force must also, Yukawa reasoned, be carried by a particle. The particle became known as the "meson" (which has to be intermediate between those of the electron and the proton, hence the name) by applying the quantum rules to the nucleus. Like the photon, mesons are bosons, but with one unit of spin, not zero; unlike photons they have very short lifetimes, which is why they are only seen outside the nucleus under special conditions. In due course, a family of mesons was found, not quite as Yukawa had predicted but sufficiently close to the forecast to show that the idea of nuclear particles exchanging mesons as carriers of the strong nuclear force does work in an analogous way to the exchange of photons as carriers of the electric force; Yukawa duly received the Nobel Prize in Physics in 1949.

This confirmation that nuclear forces, as well as electric forces, can be thought of purely in terms of interactions between particles is a cornerstone of the physicists' view of the world today. All forces are now regarded as interactions.

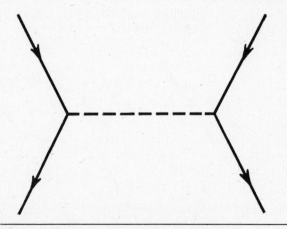

*Figure 9.9/*In a Feynman diagram, two particles interact by exchanging a third particle. In this particular case it might be two electrons, which exchange a photon and are repelled from one another.

But where do the particles that carry the interactions come from? They come from nowhere, something for nothing, in accordance with the uncertainty principle.

The uncertainty principle applies to the complementary properties of time and energy, as well as to position/momentum. The less uncertainty there is about the energy involved in an event at the particle level, the more uncertainty there is about the time of the event, and vice versa. An electron does not exist in isolation, because it can borrow energy from the uncertainty relation, for a short enough period of time, and use it to create a photon. The snag is, almost as soon as the photon is created it has to be reabsorbed by the electron, before the world at large "notices" that energy conservation has been violated. The photons exist only for a tiny fraction of a second, less than 10^{-15} sec, but they are popping in and out of existence around the electrons all the time. It is as if each electron is surrounded by a cloud of "virtual" photons, which only need a little push, a little energy from outside, to escape and become real. An electron moving from an excited state to a lower state in an atom gives the excess energy to one of its virtual photons and lets it fly free; an electron absorbing energy traps a free photon. And the same sort of process provides the glue which holds the nucleus together.

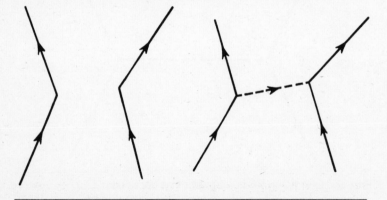

Figure 9.10/The old idea of "action at a distance" (left) is replaced by the idea of particles as carriers of the force.

Roughly speaking, since mass and energy are interchangeable, the "range" of a force is inversely proportional to the mass of the particle that provides the glue, or to the mass of the lightest particle if more than one is involved. Because photons have no mass, the range of the electromagnetic force is theoretically infinite, although it gets infinitesimally small an infinite distance away from a charged particle. Yukawa's hypothetical mesons had such a tiny range, indicated by the range of the strong nuclear force, that they had to have between 200 and 300 times the mass of the electron. As particles go, mesons are massive. The particular mesons involved in the strong nuclear interaction were found in cosmic radiation in 1946, and are called pi mesons, or pions. The uncharged or neutral pion has a mass 264 times that of the electron, and both the positive and negative pion weigh in at 273 electron masses. In round terms, they have about one seventh of the mass of a proton. Yet two protons are held together in the nucleus by exchanging, repeatedly, pions that weigh a good fraction of the protons' own weight, and without the protons losing any mass themselves. This is only possible because the protons are able to take advantage of the uncertainty principle.

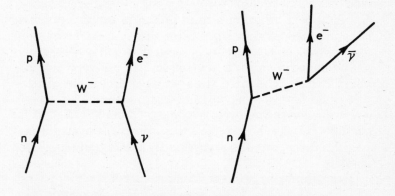

*Figure 9.11/*Two different ways of looking at the same particle interaction—just change an incoming neutrino into an outgoing antineutrino. This is the beta-decay process by which a neutron is transformed into a proton, electron, and neutrino.

A pion is created, crosses to another proton and disappears all in the twinkling of uncertainty allowed while the universe "isn't looking." Protons and neutrons—nucleons—can only exchange mesons when they are very close together, essentially when they are "touching," to use an inappropriate expression from the everyday world. Otherwise, the virtual pions cannot get across the gap during the time allowed by the uncertainty principle. So the model very neatly explains why the strong nuclear interaction is a force that has no effect on nucleons outside the nucleus, but a very powerful effect on nucleons inside the nucleus.*

So a proton is even more the center of its own cloud of activity than an electron is. As it moves on its way through space (and time) a free proton is constantly emitting and reabsorbing both virtual photons and virtual mesons. And still there is another way of looking at this phenomenon. Imagine just one proton emitting just one pion and reabsorbing it. Simple. But look at it another way. First there is one proton; then there is one proton and a pion; finally there is one proton again. Because protons are indistinguishable particles, we are at liberty to say that the first proton *disappears* and gives up its mass energy, plus a little more borrowed from the uncertainty principle, to make a pion and a new proton. Soon after, the two particles collide and disappear, creating a third proton in the process and restoring the energy balance of the universe. And why stop there? Why can't our original proton give up its energy, plus a little more, to create a *neutron* and a positively charged pion? It can. And why, then, can't a proton swap this positively charged pion with a neutron, so that it "becomes" a neutron and the neutron "becomes" a proton? That, too, is possible, just as the reverse processes involving neutrons "turning into" protons and negatively charged pions are possible.

Now things begin to get complicated, since there is no

---

*In fact, Yukawa made his calculations the other way around. He knew the range of the strong nuclear force, and this enabled him to set limits on the uncertainty of time involved in the nucleon interactions. That in turn gave him a rough idea of the energy, or mass, of the particles that carry (or mediate) the interaction.

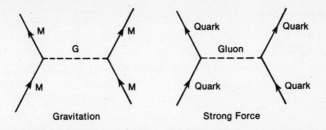

Figure 9.12/All of the fundamental forces can be represented in terms of particle exchange. In these examples, two massive particles (M) interact by the exchange of a graviton (G), and two quarks interact by the exchange of a gluon.

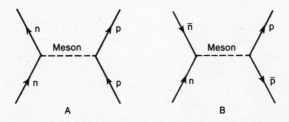

Figure 9.13/As always, the direction of time in these diagrams is an arbitrary choice. In case A, a neutrino and proton moving up the page interact by the exchange of a meson. In case B, a neutron and antineutron moving from left to right meet and annihilate to produce a meson, which in turn decays to produce a proton/antiproton pair. Such "crossed reactions" show how the concepts of force and particle become indistinguishable.

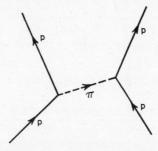

Figure 9.14/Two protons repel each other by the exchange of a pion.

reason to stop here. A pion on its own can, similarly, turn into a neutron and an antiproton for a short time, before returning to normal, and this can happen even to a virtual pion which is itself part of the Feynman pattern of a proton or a neutron. A proton proceeding quietly on its way can explode into a buzzing network of virtual particles all interacting with one another, then subside back into itself; all particles can be regarded as combinations of other particles involved in what Fritjof Capra calls "the cosmic dance." And still the story isn't over. So far, we haven't got some-

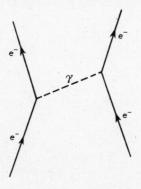

*Figure 9.15/*Two electrons interact by the exchange of a photon.

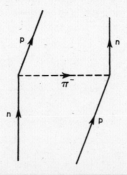

*Figure 9.16/*With the aid of a charged pion, a neutron transforms into a proton by interacting with a proton that becomes a neutron.

thing for nothing, although we have got a lot for a little. Now let's push things to extremes.

If there is an inherent uncertainty about the energy available to a particle for a short enough time, we can also say that there is an inherent uncertainty about whether or not a particle exists for a short enough time. Provided certain rules such as the conservation of electric charge and the balance between particles and antiparticles are obeyed, there is nothing to stop a whole batch of particles appearing out of nothing and then recombining with one another and disappearing before the universe at large notices the discrepancy. An electron and a positron may appear out of nothing at all, provided they disappear quickly enough; a proton and an antiproton can do the same thing. Strictly speaking, the electrons can only do the trick with the aid of a photon, and the protons with the aid of a meson, to provide the "scattering" required. A photon that doesn't exist creates a positron/electron pair, that then annihilate to produce the photon that created them in the first place—remember, the photon doesn't know the difference between future and past. Alternatively, an electron can be imagined chasing its own tail in a time eddy. First it appears, popping out of the vacuum like a rabbit out of a magician's hat, then it travels forward in time a short distance before realizing its mistake, acknowledging its own unreality, and turning around to go back from whence it came, backward through time to its starting place. There, it changes direction again, and so the loop continues, with the aid of a photon interaction—a high-energy scattering event—at each "end" of the loop.

According to our best theories of particle behavior, the vacuum is a seething mass of virtual particles in its own right, even when there are no "real" particles present. And this is not just idle tinkering with the equations, for without allowing for the effect of these vacuum fluctuations we simply do not get the right answers to problems involving scattering of particles by one another. This is powerful evidence that the theory—based directly on the uncertainty relations, remember—is correct. The virtual particles and vacuum fluctuations are as real as the rest of quantum the-

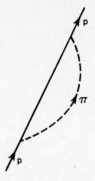

Figure 9.17/A proton can also create a "virtual" pion, provided that it is quickly reabsorbed.

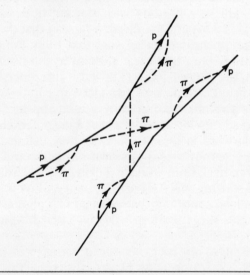

Figure 9.18/The repulsion of two protons by pion exchange is more complicated than it seemed in Figure 9.14.

ory—as real as wave/particle duality, the uncertainty principle, and action at a distance. In such a world, it hardly seems fair to call the puzzle about Schrödinger's cat a paradox at all.

SCHRÖDINGER'S CAT

The famous cat paradox first appeared in print (*Natur-wissenschaften*, vol 23 page 812) in 1935, the same year as the EPR paper. Einstein saw Schrödinger's proposal as the "prettiest way" to show that the wave representation of matter is an incomplete representation of reality,* and together with the EPR argument the cat paradox is still discussed in quantum theory today. Unlike the EPR argument, however, it has not been resolved to everyone's satisfaction.

Yet the concept behind this thought experiment is very simple. Schrödinger suggested that we should imagine a box that contains a radioactive source, a detector that records the presence of radioactive particles (a Geiger counter, perhaps), a glass bottle containing a poison such as cyanide, and a live cat. The apparatus in the box is arranged so that the detector is switched on for just long enough so that there is a fifty-fifty chance that one of the atoms in the radioactive material will decay and that the detector will record a particle. If the detector does record such an event, then the glass container is crushed and the cat dies; if not, the cat lives. We have no way of knowing the outcome of this experiment until we open the box to look inside; radioactive decay occurs entirely by chance and is unpredictable except in a statistical sense. According to the strict Copenhagen interpretation, just as in the two-hole experiment there is an equal probability that the electron goes through either hole, and the two overlapping possibilities produce a superposition of states, so in this case the equal probabilities for radioactive decay and no radioactive decay should produce a superposition of states. The whole experiment, cat and all, is governed by the rule that the superposition is "real" until we look at the experiment, and that only at the instant of observation does the wave function collapse into one of the two states. Until we look

---

*See, for example, letters 16–18 in Schrödinger's *Letters on Wave Mechanics*.

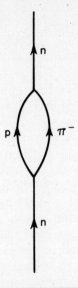

Figure 9.19/A neutron can briefly change into a proton plus a charged pion, provided the two quickly get back together.

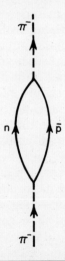

Figure 9.20/And a pion can create a virtual neutron/antiproton pair for a similarly brief interval.

inside, there is a radioactive sample that has both decayed and not decayed, a glass vessel of poison that is neither broken nor unbroken, and a cat that is both dead and alive, neither alive nor dead.

It is one thing to imagine an elementary particle such as an electron being neither here nor there but in some superposition of states, but much harder to imagine a familiar thing like a cat in this form of suspended animation. Schrödinger thought up the example to establish that there is a flaw in the strict Copenhagen interpretation, since obviously the cat cannot be both alive and dead at the same time. But is this any more "obvious" than the "fact" that an electron cannot be both a particle and a wave at the same time? Common sense has already been tested as a guide to quantum reality and been found wanting. The one sure thing we know about the quantum world is not to trust our common sense and only to believe things we can see directly or detect unambiguously with our instruments. We *don't* know what goes on inside a box unless we look.

Arguments about the cat in the box have gone on for fifty years. One school of thought says that there is no problem, because the cat is quite able to decide for itself whether it is alive or dead, and that the cat's consciousness is sufficient to trigger the collapse of the wave function. In that case, where do you draw the line? Would an ant be aware of what was going on, or a bacterium? Moving in the other direction, since this is only a thought experiment we can imagine a human volunteer taking the place of the cat in the box (the volunteer is sometimes referred to as "Wigner's friend," after Eugene Wigner, a physicist who thought deeply about variations on the cat-in-the-box experiment and who, incidentally, is Dirac's brother-in-law). The human occupant of the box is clearly a competent observer who has the quantum-mechanical ability to collapse wave functions. When we open the box, assuming we are lucky enough to find him still living, we can be quite sure that he will not report any mystic experiences but simply that the radioactive source failed to produce any particles at the alloted time. Yet still, to us outside the box the only

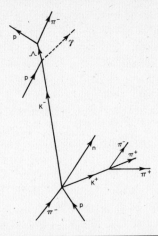

Figure 9.21/Feynman (space-time) diagram of a
genuine interaction of several particles revealed by
a bubble-chamber photograph and described by Fritjof
Capra in *The Tao of Physics*.

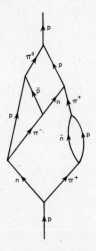

Figure 9.22/A single proton could be involved in a
network of virtual interactions like this one, from
The World of Elementary Particles, by K. Ford, Blaisdell,
New York, 1963. Such interactions go on all the time.
No particle is as lonely as it seems at first sight.

correct way to describe conditions inside the box is as a superposition of states, until we look.

The chain is endless. Imagine that we have announced the experiment in advance to an intrigued world, but to avoid press interference it has been performed behind locked doors. Even after we have opened the box and either greeted our friend or dragged the corpse out, the reporters outside don't know what is going on. To them, the whole building in which our laboratory is based is in a superposition of states. And so on, back out in an infinite regression.

But suppose we replace Wigner's friend by a computer. The computer can register the information about the radioactive decay, or lack of it. Can a computer collapse the wave

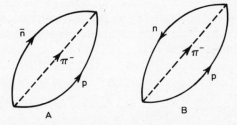

*Figure 9.23/*A proton, antineutrino, and pion can appear out of nothing at all, as a vacuum fluctuation, for a brief time before annihilating (A). The same interaction can be represented as a loop in time, with proton and neutron chasing each other around a time eddy linked by the pion (B). Both views are equally valid.

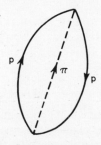

*Figure 9.24/*A proton can chase its own tail through time in the same way.

function (at least inside the box)? Why not? According to yet another point of view, what matters is not human awareness of the outcome of the experiment, or even the awareness of a living creature, but the fact that the outcome of an event at the quantum level has been recorded, or made an impact on the macroworld. The radioactive atom may be in a superposition of states, but as soon as the Geiger counter, even, has "looked" for the decay products the atom is forced into one state or the other, either decayed or not decayed.

So, unlike the EPR thought experiment, the cat-in-the-box experiment really does have paradoxical overtones. It is impossible to reconcile with the strict Copenhagen interpretation without accepting the "reality" of a dead-alive cat, and it has led Wigner and John Wheeler to consider the possibility that, because of the infinite regression of cause and effect, the whole universe may only owe its "real" existence to the fact that it is observed by intelligent beings. The most paradoxical of all the possibilities inherent in quantum theory is a direct descendant of Schrödinger's cat experiment, and it jumps off from what Wheeler calls a delayed-choice experiment.

THE PARTICIPATORY UNIVERSE

Wheeler has written many thousands of words on the meaning of quantum theory, in many different publications, over a span of four decades.* Perhaps the clearest

---

*He was born in 1911 and was just the right age to receive the full impact of the discoveries of the 1920s. Later generations have been all too willing to accept quantum theory as the received wisdom and use the quantum cookbook as the accepted rules of the game; to the older generation, relief that a consistent theory had been found, together with the natural effects of increasing age, reduced the pioneering drive. Wheeler's and Feynman's generation was inevitably the one that suffered the most soul searching about the meaning of it all, together with Einstein who, as usual, was an exception.

exposition of his concept of the "participatory universe" is in his contribution to *Some Strangeness in the Proportion,* the proceedings (edited by Harry Woolf) of a symposium held to celebrate the centenary of Einstein's birth. In that contribution (Chapter 22 of the volume) he recounts an anecdote about a time when he was playing the old game of twenty questions with a group of people at a dinner party. When it was his turn to be sent from the room so that the other guests could decide what object should be "it," he was locked out for an "unbelievably long" time, a sure sign that the collaborators were choosing a singularly difficult word or were up to some mischief. On his return, he found that the answers, from each guest in turn, came quickly at first in response to questions like "Is it an animal?" and "Is it green?" but that as the game progressed the answers took longer and longer to come, a strange procedure when all the company had, presumably, agreed on the object and the only answer required was "Yes" or "No." Why should the person questioned have to think so hard before giving a simple answer? At last, with only one question left, Wheeler guessed—"Is it a cloud?" The answer "Yes" was accompanied by a burst of laughter from the company, and he was let in on the secret.

There had been a plot *not* to agree on an object to be guessed, but that each person, when asked, must give a truthful answer concerning some real object that was in his mind, and which was *consistent with all the answers that had gone before.* As the game went on, it became as difficult for the questioned as for the questioner.

What has this to do with quantum theory? Like our concept of the real world existing out there when we are not looking at it, Wheeler imagined that there was a real answer to the object he was trying to identify. But there was not. All that was real were the answers to his questions, in the same way that the only things we know about the quantum world are the results of experiments. The cloud was, in a sense, created by the process of questioning, and in the same sense the electron is created by our process of experimental probing. The story stresses the fundamental axiom of quantum theory, that no elementary phenomenon is a phenomenon until it is a recorded phenomenon. And the process of recording can play strange tricks with our everyday concept of reality.

To make his point, Wheeler has come up with yet another thought experiment, a variation on the two-slit experiment. In this version of the game, the two slits are combined with a lens to focus light passing through the system, and the standard screen is replaced by another lens that can cause photons coming from each of the two slits to diverge. A photon that passes through one slit goes through the second screen and is deflected by the second lens off to a detector on the left; a photon passing through the other slit goes to a detector on the right. With this experimental setup, we know which slit each photon went through, as surely as in the version where we watch each slit to see if the photon passes. Just as in that case, if we allow one photon at a time to pass through the apparatus we unambiguously identify the path it follows and there is no interference because there is no superposition of states.

Now modify the apparatus again. Cover the second lens with a photographic film arranged in strips like a venetian blind. The strips can be closed to make a complete screen, preventing the photons from passing through the lens and being deflected. Or the strips can be opened, allowing the photons to pass as before. Now, when the strips are closed the photons arrive at a screen just as in the classic two-hole experiment. We have no way of telling which hole each one went through, and we have an interference pattern as if each individual photon went through both slits at once. Now comes the trick. With this setup, we don't have to decide whether to open or close the strips until the photon has already passed the two holes. We can wait until after the photon has passed the two slits, and *then* decide whether to create an experiment in which it has gone through one hole alone or through "both at once." In this delayed-choice experiment, something we do *now* has an irretrievable influence on what we can say about the past. History, for one photon at least, depends upon how we choose to make a measurement.

Philosophers have long pondered the fact that history

has no meaning—the past has no existence—except in the way it is recorded in the present. Wheeler's delayed-choice experiment fleshes out this abstract concept into solid, practical terms. "We have no more right to say 'what the photon is doing'—until it is registered—than we do to say 'what word is in the room'—until the game of question and response is terminated" (*Some Strangeness*, page 358).

How far can this concept be pushed? The happy quantum cooks, constructing their computers and manipulating genetic material, will tell you that it is all philosophical speculation, and that it doesn't mean anything in the everyday, macroscopic world. But everything in the macroscopic world is made of particles that obey the quantum rules. Everything we call real is made of things that cannot be regarded as real; "what choice do we have but to say that in some way, yet to be discovered, they all must be built upon the statistics of billions upon billions of such acts of observer-participation?"

*Figure 9.25/*Wheeler's delayed-choice
double-slit experiment (see text).

Never afraid to make the grand intuitive leap (remember his vision of the single electron weaving its way
through space and time), Wheeler goes on to consider the
whole universe as a participatory, self-excited circuit. Starting from the Big Bang, the universe expands and cools;
after thousands of millions of years it produces beings capable of observing the universe, and "acts of observer-
participancy—via the mechanism of the delayed-choice experiment—in turn give tangible 'reality' to the universe not
only now but back to the beginning." By observing the photons of the cosmic background radiation, the echo of the
Big Bang, we may be creating the Big Bang and the universe. If Wheeler is correct, Feynman was even closer to
the truth than he realized when he said that the two-hole
experiment "contains the *only* mystery."

Following Wheeler, we have wandered into the realms
of metaphysics, and I can imagine many readers thinking
that since all this depends upon hypothetical thought experiments, you can play any game you like and it doesn't

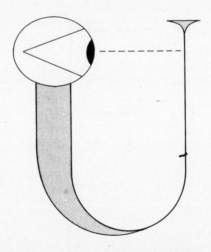

*Figure 9.26/*The whole universe can be thought of as
a delayed-choice experiment in which the existence
of observers who notice what is going on is what
imparts tangible reality to the origin of everything.

really matter which interpretation of reality you subscribe to. What we need is some solid evidence from real experiments on which to base a judgment of the best choice of interpretation from the many metaphysical options open to us. And that solid evidence is just what was supplied by the Aspect experiment in the early 1980s—the proof that quantum weirdness is not only real but observable and measurable.

CHAPTER TEN

THE PROOF
OF THE PUDDING

The direct, experimental proof of the paradoxical reality of the quantum world comes from modern versions of the EPR thought experiment. The modern experiments don't involve measurements of the position and momentum of particles, but of spin and polarization—a property of light that is in some ways analogous to the spin of a material particle. David Bohm, of Birkbeck College in London, introduced the idea of spin measurements in a new version of the EPR thought experiment in 1952, but it wasn't until the 1960s that anybody seriously considered actually performing experiments to test the predictions of quantum theory in such situations. The conceptual breakthrough came in a 1964 paper by John Bell, a physicist working at CERN, the European research center near Geneva.* But to understand the experiments, we need first to step back a little from that key paper and make sure we have a clear idea what "spin" and "polarization" mean.

---

*J. S. Bell, *Physics*, volume 1 page 195, 1964.

THE SPIN PARADOX

Fortunately, many of the peculiarities of the spin of a particle like an electron can be ignored in these experiments. It doesn't matter that the particle has to "turn around" twice before it shows the same face to us again. What does matter is that the spin of a particle defines a direction in space, up and down, analogous to the way the spin of the earth defines the direction of the north-south axis. Compared with a uniform magnetic field, an electron can only line up in one of two possible states, parallel to the field or antiparallel, "up" or "down" according to an arbitrary convention. Bohm's variation on the EPR argument starts out with a pair of protons associated with one another in a configuration called the singlet state. The total angular momentum of such a pair of protons is always zero, and we can then imagine the molecule splitting into its two component particles which depart in opposite directions. Each of those two protons can have angular momentum, and spin, but they must have equal and opposite amounts of spin, to make sure that the total for the pair is still zero, as it was when they were together.*

This is a simple prediction where quantum theory and classical mechanics both agree. If you know the spin of one particle of the pair, you know the spin of the other since the total is zero. But how do you measure the spin of one particle? In the classical world, the measurement is simple. Because we are dealing with particles in a three-dimensional world, we have to measure three directions of spin. The three components added together (using the rules of vector arithmetic, which I won't go into here) then give the total spin. But in the quantum world, the situation is very different. First, by measuring one component of spin you

---

*In this example I am following the very clear and detailed description of the Bell experiment by Bernard d'Espagnat in "The Quantum Theory and Reality," *Scientific American Offprint,* number 3066. My version is, however, highly simplified and d'Espagnat's article includes much more detail.

change the other components; the spin vectors are complementary properties and cannot be measured simultaneously any more than position and momentum can be
measured simultaneously. Secondly, the spin of a particle
such as an electron or proton is itself quantized. If you measure the spin in any direction, you can only get the answer
up or down, sometimes written as $+1$ or -1. Measure the
spin in one direction, which we might call the z-axis, and
you may get the answer $+1$ (there is an exactly fifty-fifty
chance of this outcome to the experiment). Now measure
the spin in a different direction, say along the y-axis. Whatever answer you get, go back and remeasure the spin in the
first direction you looked, the one you already "know." Repeat the experiment often, and look at the answers you get.
It turns out that no matter that you measured the spin of
the particle in the z direction and knew it was "up" before
you measured the spin in the y direction, after the y measurement you only get an answer "up" for the repeated z
measurement half the time. Measuring the complementary
spin vector has restored the quantum uncertainty of the
state you previously measured.*

So what happens when we try to measure the spin of
one of our two separating particles? Considered in isolation,
each particle can be thought of as undergoing random fluctuations in its spin components that will confuse any attempt to measure the total spin of either particle. But taken
together, the two particles must have exactly equal and opposite spin. So the random fluctuations in spin of one particle must be matched by balancing, equal, and opposite
"random" fluctuations in the spin components of the other
particle, far away. As in the original EPR argument, the particles are connected by action at a distance. Einstein regarded this "ghostly" nonlocality as absurd, implying a flaw
in quantum theory. John Bell showed how experiments

---

*Perhaps you think the uncertainty ought to be $\hbar$? It is. The fundamental unit
of spin is $\frac{1}{2}\hbar$ as Dirac established, and this is what we mean by the shorthand
"$+1$ unit of spin." The difference between $+1$ unit and -1 unit is the difference between plus and minus $\frac{1}{2}\hbar$ which, of course, is just $\hbar$. But in the
experiments discussed here, the only thing that matters is the *direction* of the
spin.

could be set up to measure this ghostly nonlocality and prove quantum theory correct.

THE POLARIZATION PUZZLE

Most of the experiments carried out so far to make this test have involved polarization of photons rather than spin of material particles, but the principle is the same. Polarization is a property that defines a direction in space associated with a photon, or a beam of photons, just as spin defines a direction in space associated with a material particle. Polaroid sunglasses work by blocking out all the photons that do not have a certain polarization, making the scene being viewed by the wearer of the glasses darker. Imagine the sunglasses as made up of a series of slats, like a venetian blind, and the photons as carrying long spears. All the photons holding their spears slantwise across their chests can slip through the slats and be seen by your eyes; all the photons marching with their spears upright cannot get through the narrow slots and are blocked out. Ordinary light contains all kinds of polarization—photons with their

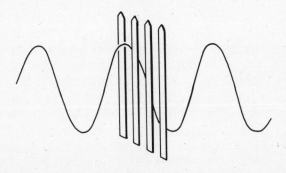

*Figure 10.1/*Vertically polarized waves slip through a "picket fence."

spears held at all different angles. There is also a kind of polarization called circular polarization, where the direction of the polarization changes as the photon advances, as if, to mix my analogies, it is represented by the orientation of the baton being twirled by a drum majorette at the head of a procession. This comes in two varieties, right-handed and left-handed, and can also be used in tests of the accuracy of the quantum view of the world. Plane polarized light, in which all the photons are holding their spears at the same angle, can be produced by reflection, under the right circumstances, or by passing the light through a substance,

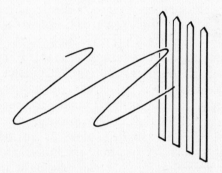

*Figure 10.2/*Horizontally polarized waves are blocked.

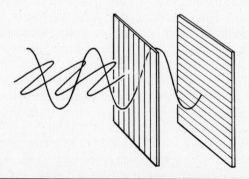

*Figure 10.3/*Crossed polarizers stop all waves.

like a polaroid lens, which only allows a certain polarization to pass. And plane polarized light shows, once again, the rules of quantum uncertainty at work.

Like the spin of a particle at the quantum level, the polarization of a photon in one direction or another is a "yes/no" property. Either it is polarized in a certain direction— vertically, perhaps—or it is not. So photons that pass through one venetian blind ought to be blocked by another one at right angles to it. If the first polarizer is like a horizontally slotted venetian blind, the second is like a vertically slitted picket fence. Sure enough, when two pieces of polarizing material are "crossed" in this way, no light gets through. But suppose the second sheet of polaroid is held so that its "slots" make a 45° angle with those of the first? The photons that arrive at this second polarizer are all 45° out of true, and on a classical picture they ought not to get through. The quantum picture is different. From that perspective, each photon has a 50 percent chance of getting through the misaligned polarizer, and half the incident photons duly get through. Now comes the really strange thing. Those photons that do get through have, in effect, been twisted. They are polarized at 45° to the original polarizer, so what happens if they now encounter another polarizer at right angles to the first? Since a right angle is 90°, they must be at 45° to this polarizer, too. So, as before, half of them get through.

With two crossed polarizers, then, no light at all gets through. But if you place a third polarizer in between the crossed pair, at 45° to them both, one quarter of the light that gets through the first polarizer also gets right through both the other two. It is as if we had two fences that together worked 100 percent at keeping stray animals off our property, but being cautious we decided to build a third fence in between the two, just to play safe. To our surprise, we now find that some of the strays that were kept out by the double fence have no difficulty walking right through the triple fence as if it weren't there. By changing the experiment we change the nature of quantum reality. In effect, by using polarizers at different angles we are measuring different vector components of polarization, and each new measurement destroys the validity of the information we got out of all the previous measurements.

This immediately introduces a new variation on the EPR theme. Instead of material particles, we deal in photons, but the basic experiment is as before. Now we imagine some atomic process that produces two photons traveling in opposite directions. There are many real processes that do this, and in such processes there is always a correlation between the polarizations of the two photons. They must either be polarized in the same way, or in some sense in opposite ways. For simplicity, in our thought ex-

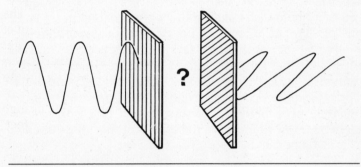

*Figure 10.4/*Two polarizers at 45° pass half the waves that get through the first!

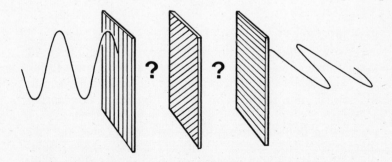

Figure 10.5/Three such polarizers pass one quarter of the waves that get through the first—even though *none* get through if the middle polarizer is removed.

periment we imagine that the two polarizations must be the same. Long after the two photons have left their birthplace, we decide to measure the polarization of one of them. We are free to choose, entirely arbitrarily, in which direction we line up our piece of polarizing material, and once we do so there is a certain chance that the photon will pass through it. We know afterward whether the photon is polarized "up" or "down" for that chosen direction in space, and we know that, far away across space, the other photon is polarized the same way. But how does the other photon know? How can it take care to orientate itself so that it will pass the same test that the first photon passes and fail the same test that the first photon fails? By measuring the polarization of the first photon we collapse the wave function, not just of one photon but of another, far away, *at the same time*.

For all its peculiarities, however, this is no more than the puzzle that Einstein and his colleagues drew to the attention of scientists in the 1930s. One real experiment is worth far more than half a century of discussion about the meaning of a thought experiment, and Bell gave experimenters a way to measure the effects of this ghostly action at a distance.

THE BELL TEST

Bernard d'Espagnat, of the University of Paris-South, is a theorist who, like David Bohm, has devoted a lot of thought to the implications of the EPR family of experiments. In his *Scientific American* article mentioned previously, and in his contribution to the volume *The Physicist's Conception of Nature,* edited by Mehra, he has spelled out the basics underlying Bell's approach to the puzzle. D'Espagnat says that our everyday view of reality is based on three fundamental assumptions. First, that there are real things that exist regardless of whether we observe them; second, that it is legitimate to draw general conclusions from consistent observations or experiments; and third, that no influence can propagate faster than the speed of light, which he calls "locality." Together, these fundamental assumptions are the basis of "local realistic" views of the world.

The Bell test starts out from a local realistic view of the world. In terms of the proton spin experiment, although the experimenter can never know the three components of spin for the same particle, he can measure any one of them he likes. If the three components are called X, Y, and Z, he finds that every time he records a value $+1$ for the X spin of one proton, he finds a value -1 for the X spin of its counterpart, and so on. But he is allowed to measure the X spin of one proton and the Y (or Z, but not both) spin of its counterpart, and in that way it ought to be possible to get information about *both* the X and Y spins of each of the pair.

Even in principle, this is far from easy, and involves measuring the spins of lots of pairs of protons at random and discarding the ones that just happen to end up measuring the same spin vector in both members of the pair. But it

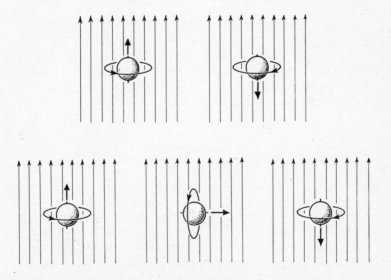

Figure 10.6/Particles with half-integer spin can only line up parallel or antiparallel to a magnetic field. Particles with integer spin are also able to align across the field.

can be done, and this gives the experimenter, in principle, sets of results in which pairs of spins have been identified for pairs of protons in sets that can be written XY, XZ, and YZ. What Bell showed in his classic 1964 paper was that if such an experiment is carried out, then according to the local realistic views of the world the number of pairs for which the X and Y components both have positive spin $(X+Y+)$ must always be less than the combined total of the pairs in which the XZ and YZ measurements all show a positive value of spin $(X+Z+ + Y+Z+)$. The calculation follows directly from the obvious fact that if a measurement shows that a particular proton has spin $X+$ and $Y-$, for example, then its total spin state must be either $X+Y-Z+$ or $X+Y-Z-$. The rest follows from a mathematically simple argument based on the theory of sets. But in quantum mechanics the mathematical rules are different, and if they are carried through correctly they come up with the opposite prediction, that the number of $X+Y+$ pairs is *more*, not less, than the number of $X+Z+$ and $Y+Z+$ pairs combined.

Because the calculation was originally expressed starting out from the local realistic view of the world, the conventional phrasing is that the *first* inequality is called "Bell's inequality," and that if Bell's inequality is *violated* then the local realistic view of the world is false, but quantum theory has passed another test.

THE PROOF

The test ought to apply equally well for the spin measurements of material particles, which are very difficult to carry out, or for measurements of the polarization of photons, which are easier to carry out, though still difficult. Because photons have zero rest mass, move at the speed of light, and have no means of distinguishing time, however, some physicists are uneasy about the experiments that involve photons. It is not really clear what the concept of locality means for a photon. So although most of the tests of Bell's inequality carried out so far have involved measurements of the polarization of photons, it is crucially important that the only test so far carried out actually using measurements of proton spins does give results that violate Bell's inequality and therefore support the quantum view of the world.

This was not the first test of Bell's inequality, but it was reported in 1976 by a team at the Saclay Nuclear Research Centre in France. The experiment very closely follows the original thought experiment, and involves shooting low-energy protons at a target that contains a lot of hydrogen atoms. When a proton strikes the nucleus of a hydrogen atom—which is another proton—the two particles interact through the singlet state and their spin components can be measured. The difficulties of making the measurements are immense. Only some of the protons are recorded by the detectors, and, unlike the ideal world of the thought experiment, even when measurements are made it is not always possible to record the spin components unambiguously. Nevertheless, the results of this French experiment clearly demonstrate that local realistic views of the world are false.

The first tests of Bell's inequality were carried out at the University of California, Berkeley, using photons, and were reported in 1972. By 1975 six such tests had been carried out, and four of them produced results that violated Bell's inequality. Whatever the doubts about the meaning of locality for photons, this is striking futher evidence in favor of quantum mechanics, especially since the experiments used two fundamentally different techniques. In the earliest photon version of the experiment, the photons came from atoms of either calcium or mercury, which can be excited by laser light into a chosen energetic state.* The route

---

*Even here we get a scent of the kind of problems that puzzled Bohr for so long. The only real things are the results of our experiments, and the way we make measurements influences what we measure. Here, in the 1980s, are physicists using as an everyday tool of their trade a laser beam whose job is simply to pump atoms into an excited state. We can only use this tool because we know about excited states and have the quantum cookbook to hand, but the whole purpose of our experiment is to verify the accuracy of quantum mechanics, the theory we used to write the quantum cookbook! I'm not sug-

back from this excited state to the ground state involves an electron in two transitions, first to another, but lower, excited state, and then on to the ground state; and each transition produces a photon. For the transitions chosen in these experiments, the two photons are produced with correlated polarizations. The photons from the cascade can then be analyzed, using photon counters that are placed behind polarizing filters.

In the mid-1970s, experimenters carried out the first measurements using another variation on the theme. In these experiments, the photons are gamma rays produced when an electron and a positron annihilate. Again, the polarizations of the two photons must be correlated, and again the balance of evidence is that, when you try to measure those polarizations, the results you get violate Bell's inequality.

So out of the first seven tests of Bell's inequality five came out in favor of quantum mechanics. In his *Scientific American* article, d'Espagnat stresses that this is even stronger evidence in favor of quantum theory than it seems at first sight. Because of the nature of the experiments and the difficulties of operating them, "a great variety of systematic flaws in the design of an experiment could destroy the evidence of a real correlation . . . on the other hand, it is hard to imagine an experimental error that could create a false correlation in five independent experiments. What is more, the results of those experiments not only violate Bell's inequality but also violate it precisely as quantum mechanics predicts."

Since the mid-1970s, even more tests have been carried out, designed to remove any remaining loopholes in the ex-

---

gesting that the experiments are therefore wrong. We can imagine other ways to excite atoms before we make the measurements, and other versions of the experiment do give the same result. But just as the everyday conceptions of previous generations of physicists were colored by their use of, say, spring balances and meter rules, so the present generation is affected, far more than it sometimes realizes, by the quantum tools of the trade.

Philosophers may care to take up the question of what the results of the Bell experiment really mean if we are using quantum processes in order to set up the experiment. I am happy to stick with Bohr—what we see is what we get; nothing else is real.

perimental design. The bits of the apparatus need to be placed far enough apart so that any "signal" between the detectors that might produce a spurious correlation would have to travel faster than light. That was done, and still Bell's inequality was violated. Or perhaps the correlation occurs because the photons "know," even as they are being created, what kind of experimental apparatus has been set up to trap them. That could happen, without the need for faster than light signals, if the apparatus has been set up in advance, and has established an overall wave function that affects the photon at birth. The ultimate test, so far, of Bell's inequality therefore involves changing the structure of the experiment while the photons are in flight, very much in the way that the double-slit experiment can be changed while the photon is in flight in John Wheeler's thought experiment. This is the experiment in which Alain Aspect's team at the University of Paris-South closed the last major loophole for the local realistic theories in 1982.

Aspect and his colleagues had already carried out tests of the inequality using photons from a cascade process, and found that the inequality was violated. Their improvement involved the use of a switch that changes the direction of a beam of light passing through it. The beam can be directed toward either one of two polarizing filters, each measuring a different direction of polarization, and each with its own photon detector behind it. The direction of the light beam passing through this switch can be changed with extraordinary rapidity, every 10 nanoseconds (10 thousand-millionths of a second, 10×10^{-9} s), by an automatic device that generates a pseudorandom signal. Because it takes 20 nanoseconds for a photon to travel from the atom in which it is born in the heart of the experiment to the detector itself, there is no way in which any information about the experimental setup can travel from one part of the apparatus to the other and affect the outcome of any measurement—unless such an influence travels faster than light.

WHAT DOES IT MEAN?

The experiment is very nearly perfect; even though the switching of the light beam is not quite at random, it does change independently for each of the two photon beams. The only real loophole that remains is that most of the photons produced are not detected at all, because the detectors themselves are very inefficient. It is still possible to argue that only the photons that violate Bell's inequality are detected, and that the others would obey the inequality if only we could detect them. But no experiments designed to test this unlikely possibility are yet contemplated, and it does seem like the height of desperation to make that argument. Following the announcement of the results from Aspect's team just before Christmas 1982,* nobody seriously doubts that the Bell test confirms the predictions of quantum theory. In fact, the results of this experiment, the best that can be achieved with present-day techniques, violate the inequalities to a greater extent than those of any of the previous tests and agree very well with the predictions of quantum mechanics. As d'Espagnat has said, "Experiments have recently been carried out that would have forced Einstein to change *his* conception of nature on a point he always considered essential . . . we may safely say that non-separability is now one of the most certain general concepts in physics."†

This does not mean that there is any likelihood of being able to send messages faster than the speed of light. There is no prospect of conveying useful information in this way, because there is no way of linking one event that *causes* another event to the event it causes through this process. It is an essential feature of the effect that it only applies to events that have a *common* cause—the annihilation of a positron/electron pair; the return of an electron to the

---

*Physical Review Letters, volume 49, page 1804.
†The Physicist's Conception of Nature, ed. J. Mehra, page 734.

ground state; the separation of a pair of protons from the singlet state. You may imagine two detectors situated far apart in space, with photons from some such central source flying off to each of them, and you may imagine some subtle technique for changing the polarization of one beam of photons, so that an observer far away at the second detector sees changes in the polarization of the other beam. But what sort of signal is it that is being changed? The original polarizations, or spins, of the particles in the beam are a result of random quantum processes, and carry no information in themselves. All that the observer will be seeing is a different random pattern from the random pattern that he would see without the cunning manipulations of the first polarizer! Since there is no information in a random pattern, it would be totally useless. The information is contained in the difference between the two random patterns, but the first pattern never existed in the real world, and there is no way to extract the information.

But don't be too disappointed, for the Aspect experiment and its predecessors do indeed make for a very different world view from that of our everyday common sense. They tell us that particles that were once together in an interaction remain in some sense parts of a single system, which responds together to further interactions. Virtually everything we see and touch and feel is made up of collections of particles that have been involved in interactions with other particles right back through time, to the Big Bang in which the universe as we know it came into being. The atoms in my body are made of particles that once jostled in close proximity in the cosmic fireball with particles that are now part of a distant star, and particles that form the body of some living creature on some distant, undiscovered planet. Indeed, the particles that make up my body once jostled in close proximity and interacted with the particles that now make up your body. We are as much parts of a single system as the two photons flying out of the heart of the Aspect experiment.

Theorists such as d'Espagnat and David Bohm argue that we must accept that, literally, everything is connected to everything else, and only a holistic approach to the universe is likely to explain phenomena such as human consciousness.

It is too early yet for the physicists and philosophers groping toward such a new picture of consciousness and the universe to have produced a satisfactory outline of its likely shape, and speculative discussion of the many possibilities touted would be out of place here. But I can provide an example from my own background, rooted in the solid traditions of physics and astronomy. One of the great puzzles of physics is the property of inertia, the resistance of an object, not to motion but to *changes* in its motion. In free space, any object keeps moving in a straight line at constant speed until it is pushed by some outside force—this was one of Newton's great discoveries. The amount of push needed to move the object depends on how much material it contains. But how does the object "know" that it is moving at constant speed in a straight line—what does it measure its velocity against? Since the time of Newton, philosophers have been well aware that the standard against which inertia seems to be measured is the frame of reference of what used to be called the "fixed stars," although we would now speak in terms of distant galaxies. The earth spinning in space, a long Foucault pendulum like the ones seen in so many science museums, an astronaut, or an atom, they all "know" what the average distribution of matter in the universe is.

Nobody knows why or how the effect works, and it has led to some intriguing if fruitless speculations. If there were just one particle in an empty universe, it couldn't have any inertia, because there would be nothing against which to measure its movement or resistance to movement. But if there were just two particles, in an otherwise empty universe, would they each have the same inertia as if they were in our universe? If we could magically remove half of the matter in our universe, would the rest still have the same inertia, or half as much? (Or twice as much?) The puzzle is as great today as it was three hundred years ago, but perhaps the death of local realistic views of the world gives us a clue. If everything that ever interacted in the Big Bang maintains its connection with everything it interacted with, then every particle in every star and galaxy that we can see "knows" about the existence of every other particle. Inertia becomes a puzzle not for cosmologists and relativists to debate, but one firmly in the arena of quantum mechanics.

Does it seem paradoxical? Richard Feynman summed up the situation succinctly in his *Lectures:* "The 'paradox' is only a conflict between reality and your feeling of what reality 'ought to be.'" Does it seem pointless, like the debate about the number of angels that can dance on a pinhead? Already, early in 1983, just a few weeks after the publication of the Aspect team's results, scientists at the University of Sussex in England were announcing the results of experiments that not only provide independent confirmation of the connectedness of things at the quantum level, but offer scope for practical applications including a new generation of computers—as much improved over present solid-state technology as the transistor radio itself is an improvement over the semaphore flag as a signaling device.

CONFIRMATION AND APPLICATIONS

The Sussex team, headed by Terry Clark, has tackled the problem of making measurements of quantum reality the other way around. Instead of trying to construct experiments that operate on the scale of normal quantum particles—the scale of atoms or smaller—they have attempted to construct "quantum particles" that are more nearly the size of conventional measuring devices. Their technique depends upon the property of superconductivity, and uses a ring of superconducting material, about half a centimeter across, in which there is a constriction at one point, a narrowing of the ring to just one ten-millionth of a square centimeter in cross section. This "weak link," invented by Brian Josephson who developed the Josephson junction, makes the ring of superconducting material act like an open-ended cylinder such as an organ pipe or a tin can with both ends removed. The Schrödinger waves describing the behavior of superconducting electrons in the ring act rather like the standing sound waves in an organ pipe, and they can be "tuned" by applying a varying electromagnetic field at radio frequencies. In effect, the electron wave around the whole of the ring replicates a single quantum particle, and by using a sensitive radio-frequency detector the team is able to observe the effects of a quantum transition of the electron wave in the ring. It is, for all practical purposes, as if they had a single quantum particle half a centimeter across with which to work—a similar, but even more dramatic, example to the little bucket of superfluid helium mentioned earlier.

The experiment provides direct measurements of single quantum transitions, and it also provides further clear evidence of nonlocality. Because the electrons in the superconductor act like one boson, the Schrödinger wave that makes a quantum transition is spread out around the whole ring. The whole of this pseudoboson makes the transition at the same time. It is *not* observed that one side of the ring makes the transition first, and that the other side only catches up when a signal moving at the speed of light has had time to travel around the ring and influence the rest of the "particle." In some ways, this experiment is even more powerful than the Aspect test of Bell's inequality. That test depends upon arguments which, although mathematically unambiguous, are not easy for the layman to follow. It is much easier to grasp the concept of a single "particle" that is half a centimeter across and yet behaves like a single quantum particle, and that responds, in its entirety, instantaneously to any prodding it receives from the outside.

Already, Clark and his colleagues are working on the next logical development. They hope to construct a larger "macroatom," perhaps in the form of a straight cylinder 6 meters long. If this device responds to outside stimulation in the way expected, then there may indeed be a crack opened in the door that leads to faster-than-light communication. A detector at one end of the cylinder, measuring its quantum state, will respond instantly to a change in the quantum state triggered by a prod at the other end of the cylinder. This is still not much use for conventional signaling—we couldn't build a macroatom reaching from here to the moon, say, and use it to eliminate that annoying lag in communications between lunar explorers and ground control here on earth. But it would have direct, practical applications.

In the most advanced modern computers, one of the key limiting factors on performance is the speed with which electrons can get around the circuitry from one component to another. The time delays involved are small, down in the nanosecond range, but very significant. The prospect of communicating instantaneously across great distances is in no way made more likely by the Sussex experiments, but the prospect of building computers in which all the components respond instantaneously to a change in the state of one component *is* brought within the realm of possibility. It is this prospect that has encouraged Terry Clark to make the claim that "when its rules are translated into circuit hardware it will make the already amazing electronics of the twentieth century look like semaphore by comparison."*

So not only is the Copenhagen interpretation fully vindicated for all practical purposes by the experiments, it looks as if there are developments in store as far beyond those that quantum mechanics has already given us as those developments are beyond classical devices. But still the Copenhagen interpretation is intellectually unsatisfying. What happens to all those ghostly quantum worlds that collapse with their wave functions when we make a measurement of a subatomic system? How can an overlapping

---

*In the *Guardian*, 6 January, 1983. While I was preparing this chapter for the printer, news of a similar development along these lines came from the Bell Laboratories, where researchers are using Josephson junction technology to develop new, fast "switches" for computer circuitry. These switches use only "conventional" Josephson junctions, and can already operate ten times faster than standard computer circuitry. That development is likely to continue to make headlines, and achieve practical applications, in the near future. But don't be confused—the developments Clark is talking about are more remote, and may not be applied before the end of this century, but are potentially a far greater leap forward.

reality, no more and no less real than the one we eventually measure, simply disappear when the measurement is made? The best answer is that the alternative realities do not disappear, and that Schrödinger's cat really is both alive and dead at the same time, but in two or more different worlds. The Copenhagen interpretation, and its practical implications, are fully contained within a more complete view of reality, the many-worlds interpretation.

CHAPTER ELEVEN

MANY WORLDS

Until now, I have tried not to take sides in this book, but to present the story of the quantum in all its aspects, and let the story speak for itself. Now the time has come to stand up and be counted. In this final chapter I abandon any pretense of impartiality and present the interpretation of quantum mechanics that I find the most satisfactory and pleasing. This is not the majority view; most of the physicists who bother to think about such things at all are happy with the collapsing wave functions of the Copenhagen interpretation. But it is a respectable minority view, and it has the merit of including the Copenhagen interpretation within itself. The uncomfortable feature that has prevented this improved interpretation from taking the world of physics by storm is that it implies the existence of many other worlds—possibly an infinite number of them—existing in some way sideways across time from our reality, parallel to our own universe but forever cut off from it.

WHO OBSERVES THE OBSERVERS?

This many-worlds interpretation of quantum mechanics originated in the work of Hugh Everett, a graduate student at Princeton University in the 1950s. Puzzling over the peculiar way in which the Copenhagen interpretation requires wave functions to collapse magically when observed, he discussed alternatives with many people, including John Wheeler who encouraged Everett to develop his alternative approach as his PhD thesis. This alternative view starts from a very simple question that is the logical culmination of considering the successive collapses of the wave function implied when I carry out an experiment in a closed room, then come out and tell you the result, which you tell a friend in New York, who reports it to someone else, and so on. At each step, the wave function becomes more complex, and embraces more of the "real world." But at each stage the alternatives remain equally valid, overlapping realities, until the news of the outcome of the experiment arrives. We can imagine the news spreading across the whole universe in this way, until the whole universe is in a state of overlapping wave functions, alternative realities that only collapse into one world when observed. But who observes the universe?

By definition, the universe is self-contained. It includes everything, so there is no outside observer who notices the existence of the universe and thereby collapses its complex web of interacting alternative realities into one wave function. Wheeler's idea of consciousness—ourselves—as the crucial observer operating through reverse causality back to the Big Bang is one way out of this dilemma, but it involves a circular argument as puzzling as the puzzle it is supposed to eliminate. I would prefer even the solipsist argument, that there is only one observer in the universe, myself, and that my observations are the all-important factor that crystalizes reality out of the web of quantum possibilities—but extreme solipsism is a deeply unsatisfying philosophy for someone whose own contribution to the world is writing books to be read by other people. Everett's many worlds interpretation is another, more satisfying and more complete possibility.

Everett's interpretation is that the overlapping wave functions of the whole universe, the alternative realities that interact to produce measurable interference at the quantum level, do not collapse. All of them are equally real, and exist in their own parts of "superspace" (and super-time). What happens when we make a measurement at the quantum level is that we are forced by the process of observation to select one of these alternatives, which becomes part of what we see as the "real" world; the act of observation cuts the ties that bind alternative realities together, and allows them to go on their own separate ways through superspace, each alternative reality containing its own observer who has made the same observation but got a different quantum "answer" and thinks that he has "collapsed the wave function" into one single quantum alternative.

SCHRÖDINGER'S CATS

It is hard to grasp what this means when we talk about the collapse of the wave function of the whole universe, but much easier to see why Everett's approach represents a step forward if we look at a more homely example. Our search for the real cat hidden inside Schrödinger's paradoxical box has, at last, come to an end, for that box provides just the example I need to demonstrate the power of the many-worlds interpretation of quantum mechanics. The surprise is that the trail leads not to one real cat, but two.

The equations of quantum mechanics tell us that inside the box of Schrödinger's famous thought experiment there are versions of a "live cat" and "dead cat" wave function that are equally real. The conventional, Copenhagen interpretation looks at these possibilities from a different perspective, and says, in effect, that both wave functions are equally *unreal,* and that only one of them crystalizes as reality when we look inside the box. Everett's interpretation accepts the quantum equations entirely at face value and says that both cats are real. There is a live cat, and there is a dead cat; but they are located in different worlds. It is not that the radioactive atom inside the box either did or didn't decay, but that it did both. Faced with a decision, the whole world—the universe—split into two versions of itself, identical in all respects except that in one version the atom decayed and the cat died, while in the other the atom did not decay and the cat lived. It sounds like science fiction, but it goes far deeper than any science fiction, and it is based on impeccable mathematical equations, a consistent and logical consequence of taking quantum mechanics literally.

BEYOND
SCIENCE FICTION

The importance of Everett's work, published in 1957, is that he took this seemingly outrageous idea and put it on a secure mathematical foundation using the established rules of quantum theory. It is one thing to speculate about the nature of the universe, but quite another to develop those speculations into a complete, self-consistent theory of reality. Indeed, Everett was not the first person to speculate in this way, although he seems to have produced his ideas totally independently of any earlier suggestions about multiple realities and parallel worlds. Most of those earlier speculations—and many more since 1957—have indeed appeared in the pages of science fiction. The earliest version I have been able to trace is in Jack Williamson's *The Legion of Time,* first published as a magazine serial in 1938.*

---

*Timewarps, an earlier book of mine, is all about parallel worlds, but includes only the minimum necessary discussion of quantum theory.

Many SF stories are set in "parallel" realities, where the South won the American Civil War, or the Spanish Armada succeeded in conquering England, and so on. Some describe the adventures of a hero who travels sideways through time from one alternative reality to another; a few describe, with suitable gobbledygook language, how such an alternative world may split off from our own. Williamson's original story deals with two alternative worlds, neither of which achieves concrete reality until some key action is taken at a crucial time in the past where the courses of the two worlds diverge (there is also "conventional" time travel in this story, and the action is as circular as the argument). The idea has echoes of the collapse of a wave function as described by the conventional Copenhagen interpretation, and Williamson's familiarity with the new ideas of the 1930s is clear from the passage in which a character explains what is going on:

> With the substitution of waves of probability for concrete particles, the world lines of objects are no longer the fixed and simple paths they once were. Geodesics have an infinite proliferation of possible branches, at the whim of subatomic indeterminism.

Williamson's world is a world of ghost realities, where the heroic action takes place, with one of them collapsing and disappearing when the crucial decision is made and another of the ghosts is selected to become concrete reality. Everett's world is one of many *concrete* realities, where all the worlds are equally real, and where, alas, not even heroes can move from one reality to its neighbor. But Everett's version is science fact, not science fiction.

Let's get back to the fundamental experiment in quantum physics, the two-holes experiment. Even within the framework of the conventional Copenhagen interpretation, although few quantum cooks realize it, the interference pattern produced on the screen of that experiment when just one particle travels through the apparatus is explained as interference from two alternative realities, in one of which the particle goes through hole A, in the other of which it goes through hole B. When we look at the holes,

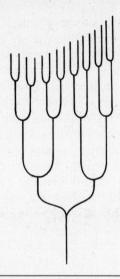

Figure 11.1/ The phrase "parallel worlds"
suggests alternative realities lying side by side in
"superspace-time." This is a false picture.

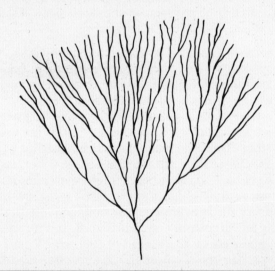

Figure 11.2/ A better picture sees the universe constantly
splitting, like a branching tree. But this is still a false image.

we find the particle only goes through one of them, and
there is no interference. But how does the particle choose
which hole to go through? On the Copenhagen interpretation, it chooses at random in accordance with the quantum
probabilities—God *does* play dice with the universe. On the
many-worlds interpretation, it doesn't choose. Faced with a
choice at the quantum level, not only the particle itself but
the entire universe splits into two versions. In one universe,
the particle goes through hole *A*, in the other it goes
through hole *B*. In each universe there is an observer who
sees the particle go through just one hole. And forever afterward the two universes are completely separate and non-
interacting—which is why there is no interference on the
screen of the experiment.

Multiply this picture by the number of quantum events
going on all the time in every region of the universe and
you get some idea of why conventional physicists balk at
the notion. And yet, as Everett established twenty-five
years ago, it is a logical, self-consistent description of quantum reality that conflicts with no experimental or observational evidence.

In spite of its impeccable mathematics, Everett's new
interpretation of quantum mechanics made scarcely a ripple as it fell into the pool of scientific knowledge in 1957. A
version of the work appeared in *Reviews of Modern Physics*,* and alongside it there appeared a paper from Wheeler
drawing attention to the importance of Everett's work†. But
the ideas remained largely ignored until they were taken up
by Bryce DeWitt, of the University of North Carolina, more
than ten years later.

It isn't clear why the idea should have taken so long to
catch on, even in the minor way that it achieved success in
the 1970s. Apart from the heavy math, Everett carefully
explained in his *Reviews of Modern Physics* paper that the
argument that the splitting of the universe into many
worlds cannot be real because we have no experience of it

---

*Volume 29, page 454.
†Volume 29, page 463.

doesn't hold water. All the separate elements of a superposition of states obey the wave equation with complete indifference as to the actuality of the other elements, and the total lack of any effect of one branch on another implies that no observer can ever be aware of the splitting process. Arguing otherwise is like arguing that the earth cannot possibly be in orbit around the sun, because if it were we would feel the motion. "In both cases," says Everett, "the theory itself predicts that our experience will be what in fact it is."

BEYOND EINSTEIN?

In the case of the many-worlds interpretation, the theory is conceptually simple, causal, and gives predictions in accord with experience. Wheeler did his best to make sure people noticed the new idea:

> It is difficult to make clear how decisively the "relative state" formulation drops classical concepts. One's initial unhappiness at this step can be matched but a few times in history: when Newton described gravity by anything so preposterous as action at a distance; when Maxwell described anything as natural as action at a distance in terms as unnatural as field theory; when Einstein denied a privileged character to any coordinate system . . . nothing quite comparable can be cited from the rest of physics except the principle in general relativity that all regular coordinate systems are equally justified.*

"Apart from Everett's concept," Wheeler concluded, "no self-consistent system of ideas is at hand to explain what one shall mean by quantizing a closed system like the universe of general relativity." Strong words, indeed; but the Everett interpretation suffers from one major defect in trying to oust the Copenhagen interpretation from its estab-

*Op. cit., page 464.

lished place in physics. The many-worlds version of quantum mechanics makes *exactly the same predictions* as the Copenhagen view when assessing the likely outcome of any experiment or observation. That is both a strength and a weakness. Since the Copenhagen interpretation has never yet been found wanting in these practical terms, any new interpretation must give the same "answers" as the Copenhagen interpretation wherever it can be tested; so the Everett interpretation passes its first test. But it only improves upon the Copenhagen view by removing the seemingly paradoxical features from double-slit experiments, or from tests of the kind invented by Einstein, Podolsky, and Rosen. From the viewpoint of all the quantum cooks, it is hard to see the difference between the two interpretations, and the natural inclination is to stick with the familiar. For anyone who has studied the EPR thought experiments, and now the various tests of Bell's inequality, however, the attraction of the Everett interpretation is much greater. In the Everett interpretation, it is not that our choice of which spin component to measure forces the spin component of another particle, far away across the universe, to magically take up a complementary state, but rather that by choosing which spin component to measure we are choosing which branch of reality we are living in. In that branch of superspace, the spin of the other particle always is complementary to the one we measure. It is *choice* that decides which of the quantum worlds we measure in our experiments, and therefore which one we inhabit, not chance. Where all possible outcomes of an experiment actually do occur, and each possible outcome is observed by its own set of observers, it is no surprise to find that what we observe is one of the possible outcomes of the experiment.

A SECOND LOOK

The many-worlds interpretation of quantum mechanics was almost studiously ignored by the physics community until DeWitt took the idea up in the late 1960s, writing about the concept himself and encouraging his student Neill Graham to develop an extension of Everett's work as his own PhD thesis. As DeWitt explained in an article in *Physics Today* in 1970*, the Everett interpretation has an immediate appeal when applied to the paradox of Schrödinger's cat. We no longer have to worry about the puzzle of a cat that is both dead and alive, neither alive nor dead. Instead, we know that in our world the box contains a cat that is either alive or dead, and that in the world next door there is another observer who has an identical box that contains a cat that is either dead or alive. But if the universe is "constantly splitting into a stupendous number of branches," then "every quantum transition taking place on every star, in every galaxy, in every remote corner of the universe is splitting our local world on earth into myriads of copies of itself."

DeWitt recalls the shock he experienced on first encountering this concept, the "idea of 10^{100} slightly imperfect copies of oneself all constantly splitting into further copies." But he was persuaded, by his own work, by Everett's thesis and by Graham's renewed study of the phenomenon. He even considers how far the splitting can actually continue to take place. In a finite universe—and there are good reasons for believing that if general relativity is a good description of reality then the universe is finite†—there must only be a finite number of "branches"

---

*Volume 23, number 9 (September 1970), page 30.
†General relativity is a theory that describes closed systems, and Einstein originally envisaged the universe as a closed, finite system. Although people talk about open, infinite universes, strictly speaking such descriptions are not properly covered by relativity theory. The way for our universe to be closed is if it contains enough matter for gravity to bend space-time around on itself, like

on the quantum tree, and superspace simply may not have enough room to accommodate the more bizarre possibilities, the fine structure of what DeWitt calls "maverick worlds," realities with strangely distorted patterns of behavior. In any case, although the strict Everett interpretation says that anything that is *possible* does occur in some version of reality, somewhere in superspace, that is not the same thing as saying anything *imaginable* can occur. We can imagine impossible things, and the real worlds could not accommodate them. In a world otherwise identical to our own, even if pigs (otherwise identical to our own pigs) had wings they would *not* be able to fly; heroes, no matter how super, cannot slip sideways through cracks in time to visit alternative realities, even though SF writers speculate about the consequences of such actions; and so on.

DeWitt's conclusion is as dramatic as the earlier conclusion of Wheeler:

> The view from where Everett, Wheeler and Graham sit is truly impressive. Yet it is a completely causal view, which even Einstein might have accepted . . . it has a better claim than most to be the natural end product of the interpretation program begun by Heisenberg in 1925.

Perhaps it is only fair, at this point, to mention that Wheeler himself has recently expressed doubts about the whole business. In response to a questioner at a symposium held to mark the centenary of Einstein's birth, he said of the many-worlds theory, "I confess that I have reluctantly had to give up my support of that point of view in the end—much as I advocated it in the beginning—because I am afraid it carries too great a load of metaphysical baggage."[*]

---

the bending of space-time around a black hole. That needs more matter than we can see in the visible galaxies, but most observations of the dynamics of the universe suggest that it is in fact in a state very close to being closed—either "just closed" or "just open." In that case, there is no observational justification for rejecting the fundamental relativistic implications that the universe is closed and finite, and there is every reason to seek the dark matter that holds it together gravitationally. Some of the basis for these ideas can be found in Wheeler's contribution to *Some Strangeness in the Proportion.*

[*]*Some Strangeness in the Proportion,* ed. Harry Woolf, pp. 385–386.

This shouldn't be read as pulling the rug from under the Everett interpretation; the fact that Einstein changed his mind about the statistical basis of quantum mechanics didn't pull the rug from under that interpretation. Nor does it mean that what Wheeler said in 1957 is no longer true. It is still true, in 1983, that apart from Everett's theory no self-consistent system of ideas is at hand to explain what is meant by quantizing the universe. But Wheeler's change of heart does show how hard many people find it to accept the many-worlds theory. Personally, I find the load of metaphysical baggage required far less disturbing than the Copenhagen interpretation of Schrödinger's experiment with the cat, or the requirement of three times as many dimensions of "phase space" as there are particles in the universe. The concepts are no stranger than other concepts that seem familiar just because they are discussed so widely, and the many-worlds interpretation offers new insights into why the universe we live in should be the way it is. The theory is far from being played out and still deserves serious attention.

BEYOND EVERETT

Cosmologists today talk quite happily about events that occurred just after the universe was born in a Big Bang, and they calculate the reactions that occurred when the age of the universe was 10^{-35} seconds or less. The reactions involve a maelstrom of particles and radiation, pair production and annihilation. The assumptions about how these reactions take place come from a mixture of theory and the observations of the way particles interact in giant accelerators, like the one run by CERN in Geneva. According to these calculations, the laws of physics determined from our puny experiments here on earth can explain in a logical and self-consistent fashion how the universe got from a state of almost infinite density into the state we see it in today. The theories even make a stab at predicting the balance between matter and antimatter in the universe, and between matter and radiation.* Everyone interested in science, however mild and passing their interest, has heard of the Big Bang theory of the origin of the universe. Theorists happily play with numbers describing events that allegedly occurred during split seconds some 15 thousand million years ago. But who today stops to think what these ideas really mean? It is absolutely mind-blowing to attempt to understand the implications of these ideas. Who can appreciate what a number like 10^{-35} of a second really means, let alone comprehend the nature of the universe when it was 10^{-35} seconds old? Scientists who deal with such bizarre extremes of nature really should not find it too difficult to stretch their minds to accommodate the concept of parallel worlds.

In fact, that felicitous-sounding expression, borrowed from science fiction, is quite inappropriate. The natural image of alternative realities is as alternative branches fanning out from a main stem and running alongside one another though superspace, like the branching lines of a complex railway junction. Like some super-superhighway, with millions of parallel lanes, the SF writers imagine all the worlds proceeding side by side through time, our near neighbors almost identical to our own world, but with the differences becoming clearer and more distinct the further we move "sideways in time." This is the image that leads naturally to speculation about the possibilitiy of changing lanes on the superhighway, slipping across into the world next door. Unfortunately, the math isn't quite like this neat picture.

Mathematicians have no trouble handling more dimensions than the familiar three space dimensions so important in our everyday lives. The whole of our world, one branch of Everett's many-worlds reality, is described mathematically in four dimensions, three of space and one of time, all at right angles to one another, and the math to describe more dimensions all at right angles to each other and to our own four is routine number juggling. This is where the alternative realities actually lie, not parallel to

---

*All these ideas are discussed in my book *Spacewarps*.

our own world but at right angles to it, *perpendicular* worlds branching off "sideways" through superspace. The picture is hard to visualize,* but it does make it easier to see why slipping sideways into an alternative reality is impossible. If you set off at right angles to our world—sideways—you would be creating a new world of your own. Indeed, on the many-worlds theory this is what happens every time the universe is faced with a quantum choice. The only way you could get in to one of the alternative realities created by such a splitting of the universe as a result of a cat-in-the-box experiment, or a two-holes experiment, would be to go back in time in our own four-dimensional reality to the time of the experiment, and then to go forward in time along the alternative branch, at right angles to our own four-dimensional world.

This might be impossible. Conventional wisdom has it that true time travel must be impossible, because of the paradoxes involved, like the one where you go back in time and kill your grandfather before your own father has been conceived. On the other hand, at the quantum level particles seem to be involved in time travel all the "time," and Frank Tipler has shown that the equations of general relativity permit time travel. It is possible to conceive of a kind of genuine travel forward and backward in time that does not permit paradoxes, and such a form of time travel depends on the reality of alternative universes. David Gerrold

---

*And if you have trouble believing this, you may be beginning to feel that the good old Schrödinger equation is more cozy and familiar. Far from it. The wave interpretation of quantum mechanics does start out from a simple wave equation familiar from other areas of physics, and for a single particle the correct quantum-mechanical description does involve a wave in three dimensions, though not in our everyday space but in something called "configuration space." Unfortunately, you need three *different* dimensions for the wave for each particle involved in the description. To describe two particles interacting, you need six dimensions; to describe a system of three particles you need nine dimensions; and so on. The wave function of the entire universe, whatever that means, is a wave function involving three times as many dimensions as there are particles in the universe. Physicists who dismiss the Everett interpretation of reality as carrying too much excess baggage conveniently forget that the wave equations they use every day can only be accepted as a good description of the universe by invoking an equally mind-boggling load of extradimensional baggage.

explored these possibilities in an entertaining Sf book *The Man Who Folded Himself,* well worth reading as a guide to the complexities and subtleties of a many-worlds reality. The point is that, taking the classic example, if you go back in time and kill your grandfather you are creating, or entering (depending on your point of view) an alternative world branching off at right angles to the world in which you started. In that "new" reality, your father, and yourself, are never born, but there is no paradox because you are still born in the "original" reality, and make the journey back through time and into an alternative branch. Go back again to undo the mischief you have done, and all you do is re-enter the original branch of reality, or at least one rather like it.

But even Gerrold does not "explain" the bizarre events that happen to his main character in terms of perpendicular realities, and as far as I know this physical explanation of the mathematics of the Everett interpretation is original—certainly a new twist to the time travel saga that the Sf writers have not, as yet, embraced. I hereby offer it to them.* The point that is worth stressing is that the alternative realities do not, on this picture, lie "alongside" ours, where they can be slipped into and out of with very little effort. Each branch of reality is at right angles to all the other branches. There may be a world in which Bonaparte's given name was Pierre, not Napoleon, but where history otherwise flowed essentially as in our own branch of reality; there may be a world in which that particular Bonaparte never existed. Both are equally remote and inaccessible from our own world. Neither can be reached except by traveling backward in time in our own world to the appropriate branching point, and then by striking out again forward in time at right angles (one of the many right angles!) to our own reality.

The concept can be extended to remove the paradoxical nature of any of the time-travel paradoxes· beloved of science fiction writers and readers, and argued over by phi-

---

*While this book was making its way to press, I wrote a short story, "Perpendicular Worlds," for *Analog,* that uses this theme.

losophers. All *possible* things do happen, in some branch of reality. The key to entering those possible realities is not travel sideways in time, but backward and then forward into another branch. Possibly the best science-fiction novel ever written makes use of the many-worlds interpretation, although I am not sure that the author, Gregory Benford, did so consciously. In his book *Timescape* the fate of a world is fundamentally altered as a result of messages that are sent back to the 1960s from the 1990s. The story is beautifully crafted, gripping and stands up in its own right even without the SF theme. But the one point I want to pick up here is that because the world is changed as a result of actions taken by the people who receive the messages from the future, the future from which those messages came does not exist for them. So where did the messages come from? You could, perhaps, make a case on the old Copenhagen interpretation for a ghost world sending back ghost messages that affect the way the wave function collapses, but you'd be hard pressed to make that argument stand up. On the other hand, in the many-worlds interpretation it is straightforward to visualize messages from one reality going back in time to a branching point where they are received by people who then move forward in time into their own, different branch of reality. Both alternative worlds exist, and communication between them is broken once the critical decisions that affect the future have been made.* *Timescape,* as well as being a good read, actually contains a "thought experiment" every bit as intriguing and relevant to the quantum mechanics debate as the EPR experiment or Schrödinger's cat. Everett himself may not have appreciated it, but a many-worlds reality is exactly the kind of reality that permits time travel. It is also the kind of

---

*There is another element here that is worth emphasizing. Even if time travel is theoretically possible, there may be insuperable practical difficulties to prevent us from sending material objects through time. But sending messages through time could be a relatively simple matter if we can find a way to make use of the particles that travel backward in time in Feynman's interpretation of reality.

reality that explains why we should be here debating such issues.

OUR SPECIAL PLACE

According to my interpretation of the many-worlds theory, the future is not determined, as far as our conscious perception of the world is concerned, but the past is. By the act of observation we have selected a "real" history out of the many realities, and once someone has seen a tree in our world it stays there even when nobody is looking at it. This applies all the way back to the Big Bang. At every junction in the quantum highway there may have been many new realities created, but the path that leads to us is clear and unambiguous. There are many routes into the future, however, and some version of "us" will follow each of them. Each version of ourselves will think it is following a unique path, and will look back on a unique past, but it is impossible to know the future since there are so many of them. We may even receive messages from the future, either by mechanical means as in *Timescape*, or, if you wish to imagine the possibility, through dreams and extrasensory perception. But those messages are unlikely to do us much good. Because there is a multiplicity of future worlds, any such messages must be expected to be confused and contradictory. If we act on them, we are more likely than not going to deflect ourselves into a branch of reality different from the one the "messages" came from, so that it is highly unlikely that they can ever "come true." The people who suggest that quantum theory offers a key to practical ESP, telepathy, and all the rest are only deluding themselves.

The picture of the universe as an unrolling Feynman diagram across which the instant "now" moves at a steady rate is an oversimplification. The real picture is of a multidimensional Feynman diagram, all the possible worlds, with "now" unrolling across all of them, up every branch and detour. The greatest question left to answer within this framework is why our perception of reality should be what it is—why should the choice of paths through the quantum maze that started out in the Big Bang and leads to us have been just the right kind of path for the appearance of intelligence in the universe?

The answer lies in an idea often referred to as the "anthropic principle." This says that the conditions that exist in our universe are the only conditions, apart from small variations, that could have allowed life like us to evolve, and so it is inevitable that any intelligent species like us should look out upon a universe like the one we see about us.* If the universe wasn't the way it is, we wouldn't be here to observe it. We can imagine the universe taking many different quantum paths forward from the Big Bang. In some of those worlds, because of differences in the quantum choices made near the beginning of the universal expansion, stars and planets never form, and life as we know it does not exist. Taking a specific example, in our universe there seems to be a preponderance of matter particles and little or no antimatter. There may be no fundamental reason for this—it may be just an accident of the way the reactions worked out during the fireball phase of the Big Bang. It is just as likely that the universe should be empty, or that it should consist chiefly of what we would call antimatter, with little or no matter present. In the empty universe, there would be no life as we know it; in the antimatter universe there could be life just like us, a kind of looking-glass world made real. The puzzle is why a world ideal for life should have appeared out of the Big Bang.

The anthropic principle says that many possible worlds may exist, and that we are an inevitable product of our kind of universe. But where are the other worlds? Are they ghosts, like the interacting worlds of the Copenhagen interpretation? Do they correspond to different life cycles of the whole universe, before the Big Bang that began time and space as we know them? Or could they be Everett's many

---

*I have discussed the anthropic principle briefly in my book *Spacewarps;* more details can be found in *The Accidental Universe,* by Paul Davies. My own *Genesis* explains in detail the Big Bang origin of the universe.

worlds, all existing at right angles to our own? It seems to me that this is by far the best explanation available today, and that the resolution of the fundamental puzzle of why we see the universe the way it is amply compensates for the load of baggage carried by the Everett interpretation. Most of the alternative quantum realities are unsuitable for life and are empty. The conditions that are just right for life are special, so when living beings look back down the quantum path that has produced themselves they see special events, branches in the quantum road that may not even be the most likely on a statistical basis, but are the ones that lead to intelligent life. The multiplicity of worlds like our own but with different histories—in which Britain still rules all of its North American colonies; or in which the North American natives colonized Europe—together make up just one small corner of a much vaster reality. It is not chance that has selected the special conditions suitable for life out of the array of quantum possibilities, but choice. All worlds are equally real, but only suitable worlds contain observers.

The success of the Aspect team's experiments to test the Bell inequality has eliminated all but two of the possible interpretations of quantum mechanics ever put forward. Either we have to accept the Copenhagen interpretation, with its ghost realities and half-dead cats, or we have to accept the Everett interpretation with its many worlds. It is, of course, conceivable that neither of the two "best buys" in the science supermarket is correct, and that both these alternatives are wrong. There may yet be another interpretation of quantum mechanical reality which resolves all of the puzzles that the Copenhagen interpretation and the Everett interpretation resolve, including the Bell Test, and which goes beyond our present understanding—in the same way, perhaps, that general relativity transcends and incorporates special relativity. But if you think this is the soft option, an easy route out of the dilemma, remember that any such "new" interpretation must explain *everything* that we have learned since Planck's great leap in the dark, and that it must explain everything as well as, or better than, the two current explanations. That is a very tall order indeed, and it is not the way of science to sit idly back and hope that someone will come up with a "better" answer to our problems. In the absence of a better answer, we have to face up to the implications of the best answer we've got. Writing in the 1980s, after more than half a century of intensive effort devoted to the puzzle of quantum reality by the best brains of the twentieth century, we have to accept that science can at present only offer these two alternative explanations of the way the world is constructed. Neither of them seems very palatable at first sight. In simple language, either nothing is real or everything is real.

The issue may never be resolved, because it may be impossible to devise an experiment to distinguish between the two interpretations, short of time travel. But it is quite clear that Max Jammer, one of the ablest of quantum philosophers, was not exaggerating when he said that "the multiuniverse theory is undoubtedly one of the most daring and most ambitious theories ever constructed in the history of science."* Quite literally, it explains everything, including the life and death of cats. As an incurable optimist, it is the interpretation of quantum mechanics that appeals most to me. All things are possible, and by our actions we choose our own paths through the many worlds of the quantum. In the world in which we live, what you see is what you get; there are no hidden variables; God doesn't play dice; and everything is real. One of the anecdotes told and retold about Niels Bohr is that when someone came to him with a wild idea purporting to resolve one of the puzzles of quantum theory in the 1920s, he replied, "Your theory is crazy, but it's not crazy enough to be true."† In my view, Everett's theory *is* crazy enough to be true, and that seems a suitable note on which to conclude our search for Schrödinger's cat.

---

*The Philosophy of Quantum Mechanics, page 517.
†Quoted by, for example, Robert Wilson, The Universe Next Door, page 156.

UNFINISHED BUSINESS

The story of the quantum as I have told it here seems neatly cut and dried, except for the semiphilosophical question of whether you prefer the Copenhagen interpretation or the many-worlds version. That is the best way to present the story in a book, but it isn't the whole truth. The story of the quantum is not yet finished, and theorists today are grappling with problems that may lead to a step forward as fundamental as the step Bohr took when he quantized the atom. Trying to write about this unfinished business is messy and unsatisfying; the accepted views of what is important and what can safely be ignored may change completely by the time the report gets into print. But to give you a flavor of how things may be progressing, I include in this epilogue an account of the unfinished aspects of the quantum story and some hints about what to watch out for in the future.

The clearest sign that there is still more to quantum theory than meets the eye comes from the branch of quantum theory that is generally regarded as the jewel in the crown, the greatest triumph of the theory. This is quantum electrodynamics, or QED for short, the theory that "explains" the electromagnetic interaction in quantum terms. QED flowered in the 1940s, and has proved so successful that it has been used as the model for a theory of the strong nuclear interaction, a theory that is in turn dubbed quantum chromodynamics, or QCD, because it involves the interactions of particles called quarks, which have properties that the theorists distinguish, whimsically, by labeling them with the names of colors. Yet QED itself suffers from a major flaw. The theory works, but only as a result of fudging the math to make it fit our observations of the world.

The problems relate to the way in which an electron in quantum theory is not the naked particle of classical theory, but is surrounded by a cloud of virtual particles. This cloud of particles must affect the mass of the electron. It is quite possible to set up the quantum equations corresponding to an electron + cloud, but whenever those equations are solved mathematically they give infinitely large "answers." Starting from the Schrödinger equation, the cornerstone of quantum cookery, the correct mathematical treatment of the electron yields infinite mass, infinite energy, and infinite charge. There is no legal mathematical way to get rid of the infinities, but it is possible to get rid of them by cheating. We know what the mass of an electron is by direct experimental measurements, and we know that this is the answer that our theory ought to give us for the mass of electron + cloud. So the theorists remove the infinities from the equations, in effect dividing one infinity by another.

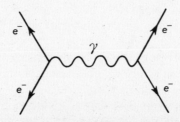

*Figure E.1/*The classic Feynman diagram of particle interactions.

Mathematically, if you divide infinity by infinity you could get any answer at all, and so they say that the answer must be the answer we want, the measured mass of the electron. This trick is called renormalization.

To get a picture of what is going on, imagine that somebody who weighs 150 pounds goes to the moon, where the gravitational force at the surface is only one sixth of the gravitational force at the surface of the earth. On a conventional bathroom scale set up on earth and brought along on the trip, the traveler's weight will be recorded as just 25 pounds, even though his body has not lost any mass. In such circumstances, it would be sensible, perhaps, to "renormalize" the bathroom scale by twiddling the control until the recorded weight showed 150 pounds. But the trick only works because we know what the weight of the traveler really is, in earth terms, and we want to keep our

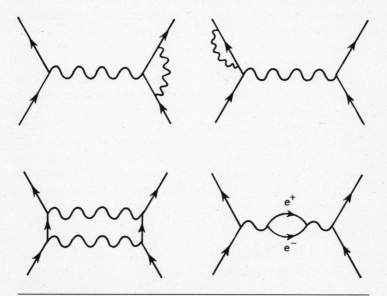

*Figure E.2/*Quantum corrections to the laws of electrodynamics arise because of the presence of virtual particles—diagrams that have closed loops. These are the situations that lead to infinities that can only be removed by the unsatisfactory trick of renormalization.

records in terms of earth weight. If the scale registered infinite weight, we could only adjust it to reality by making an infinite correction, and that is what the quantum theorists do in QED. Unfortunately, although dividing 150 by 6 unambiguously gives the answer 25, dividing $25 \times$ infinity by infinity does *not* unambiguously give the answer 25, but could give any answer at all.

Even so, the trick is immensely powerful. With the infinities canceled out, the solutions to the Schrödinger equation do everything that physicists could wish and describe perfectly even the most subtle effects of electromagnetic interactions on atomic spectra. The results are perfect, so most theorists accept QED as a good theory and don't worry about the infinities, just as the quantum cooks don't worry about the Copenhagen interpretation or the uncertainty principle. But the fact that the trick works doesn't stop it being a trick, and the one person whose opinion ought to be the most respected regarding quantum theory remains

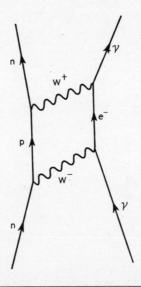

Figure E.3/Exchanging two W bosons between a neutrino and a neutron is sufficient to require an *infinite* correction to the calculation, compared with the exchange of a single boson.

deeply unhappy about it. In a lecture given in New Zealand as recently as 1975*, Paul Dirac commented:

> I must say that I am very dissatisfied with the situation, because this so-called "good theory" does involve neglecting infinities which appear in its equations, neglecting them in an arbitrary way. This is just not sensible mathematics. Sensible mathematics involves neglecting a quantity when it turns out to be small— not neglecting it just because it is infinitely great and you do not want it!

After pointing out that in his opinion "this Schrödinger equation has no solutions," Dirac concluded his lecture by stressing that there must be a *drastic* change in the theory to make it mathematically sensible. "Simple changes will not do . . . I feel that the change required will be just about as dramatic as the passage from the Bohr theory to quantum mechanics." Where can we look for such a new theory? If I had the answer to that question, I would be out winning my own Nobel Prize; but I can point you toward some of the interesting developments emerging from physics today that may ultimately satisfy even Dirac's probing investigations of what constitutes a good theory.

TWISTED SPACE-TIME

Perhaps the way to a better understanding of the nature of the universe lies in the part of the physical world that has largely been ignored in quantum theory so far. Quantum mechanics tells us a lot about material particles; it tells us scarcely anything at all about empty space. Yet as Eddington remarked more than fifty years ago in *The Nature*

---

*Directions in Physics, Chapter Two. Dirac is not alone in his concern; Banesh Hoffmann, in *The Strange Story of the Quantum*, page 213, describes renormalization as leading physics into a cul-de-sac. "The audacious juggling with infinities is extraordinarily brilliant. But its brilliance seems to illuminate a blind alley."

of the Physical World, the revolution that created our picture of solid matter as very largely empty space is more fundamental than the revolution brought about by relativity theory. Even a solid object like my desk, or this book, is actually almost all empty space. The proportion of matter to space is smaller even than the proportions of a grain of sand compared with the Albert Hall. The one thing quantum theory does seem to tell us about this neglected 99.99999 . . . percent of the universe is that it is seething with activity, a maelstrom of virtual particles. Unfortunately, the same quantum equations that yield infinite solutions in QED also tell us that the energy density of the vacuum is infinite, and renormalization has to be applied even to empty space. When the standard quantum equations are combined with those of general relativity to attempt a better description of reality, the situation is even worse—infinities still occur, but now they cannot even be renormalized. Clearly, we are barking up the wrong tree. But which tree should we be barking up?

Roger Penrose, of the University of Oxford, has gone back to basics with his attempt to make progress. He has looked at different ways to paint a geometric description of the vacuum and of particles in the vacuum, geometries involving distorted space-time and twisted pieces of space-time that we perceive as particles. For obvious reasons, the theory he has constructed is called "twistor" theory; unfortunately, not only is the math inaccessible to most people, but the theory itself is far from complete. But the concept is important—using one theory, Penrose is trying to explain both the tiny particles and the vast reaches of empty space within a solid object like this book. It may be the wrong theory, but by tackling head-on a problem that is largely ignored it does highlight one possible reason for the failures of standard theory.

There are other ways to imagine distortions of space-time down at the quantum level. By combining the constant of gravity, Planck's constant and the speed of light (the three fundamental constants of physics) it is possible to obtain a unique, basic unit of length, which might be thought of as the quantum of length, representing the

smallest region of space that can be described mean-
ingfully. It is very small indeed, about 10^{-35} meter, and it is
called the Planck length. In the same way, juggling the
fundamental constants in a different way yields one, and
only one, fundamental unit of time: the Planck time, which
is about 10^{-43} sec.* It is meaningless to talk of any interval
of time shorter than this, or any dimension of space smaller
than the Planck length.

Quantum fluctuations in the geometry of space are
completely negligible at the scale of atoms, or even elementary particles, but at this very fundamental level space itself
can be thought of as a foam of quantum fluctuations—John
Wheeler, who developed this idea, makes the comparison
between an ocean that seems flat to an aviator flying high
above it, but that seems anything but to the occupants of a
lifeboat tossing about on its stormy, ever-changing surface.† At the quantum level, space-time itself may be very
complex topologically, with "wormholes" and "bridges" con-
necting different regions of space-time; alternatively, according to a variation on the theme, empty space may be
made up of black holes, the size of the Planck length,
packed tightly together.

These are all vague, unsatisfactory, and puzzling ideas.
There are no fundamental answers here yet, but it does no
harm to be aware that our understanding of "empty space"
really is confused and uncertain, vague and unsatisfactory.
It broadens the mind to contemplate that all of the material
particles may be no more than twisted fragments of empty
space. Reasoning that if the theories we "understand"
break down, then progress is likely to come from things we
don't yet understand, it might be interesting to keep an eye
on what the quantum geometricians come up with over the
next few years. In 1983, however, the headlines over the
science news reports concerned two aspects of the good
old-fashioned particle approach to the problem.

---

*If you really want to know, the Planck length is given by the square root of
$G\hbar/c^3$ and the Planck time is the square root of $G\hbar/c^5$.
†For example, see Wheeler's contribution to Mehra's *The Physicist's Conception of Nature.*

BROKEN SYMMETRY

Symmetry is a fundamental concept in physics. The fundamental equations are time-symmetric, for example, and work equally well forward or backward in time. Other symmetries can be understood in geometrical terms. A rotating sphere, say, can be reflected in a mirror. Looking down on the top of the sphere, it may be seen rotating anticlockwise, in which case the mirror image will be rotating clockwise. Both the real sphere and the mirror image are moving in ways allowed by the laws of physics, which are symmetrical in this sense (and, of course, the mirror image sphere is also rotating in just the way that the real sphere would be seen to be rotating if time ran backward. If time is reversed *and* the mirror reflection is made, we are back where we started.). There are many other kinds of symmetry in nature. Some of these are easy to understand in everyday

*Figure E.4/*Reflection symmetry. The rotation
of the sphere in the mirror world is the same as the
time reverse of its rotation in
the real world.

language—the electron and the positron, for example, can be thought of as mirror images of each other, just as one can be thought of as a time-reversed counterpart to the other. A reversed positive charge is a negative charge. Together, these ideas of reflection in space (called a parity change, because it swaps left for right), reflection in time and reflection of charge make up one of the most powerful underlying principles in physics, the PCT theorem, which says that the laws of physics must be unaffected by changing *all three* of these to their reflected counterparts at the same time. It is the PCT theorem that is the basis of the assumption that the emission of a particle is *exactly* equivalent to the absorption of its antiparticle counterpart.

But other symmetries are much harder to come to grips with in everyday language and require the language of mathematics to be fully understood. These symmetries are crucial to an understanding of the latest news from the particle front, however, so consider a simple physical example: think of a ball balanced on a staircase. If we move the ball to another step, we change its potential energy in the gravitational field in which it is sitting. It doesn't matter how we move the ball—we can take it on a trip around the world or send it in a rocket to Mars and back before putting it on the new step. The only things that determine the change in potential energy are the heights of the two steps, the one it starts from and the one it ends on. And it doesn't matter where we choose to measure the potential energy from. We might make our measurements from the basement, and give each step a large potential energy, or we might measure from the lower of the two steps itself, in which case that step corresponds to a state with zero potential energy.* The *difference* in potential energy between the two states is still the same. This is a kind of symmetry, and because we can "regauge" our baseline from which we make the measurements, such a symmetry is called a gauge symmetry.

The same sort of thing occurs with electric forces. Max-

---

*This borrows heavily from the approach used by Paul Davies in his book *The Forces of Nature*, Cambridge University Press, 1979.

well's electromagnetism is gauge invariant as a result and QED is also a gauge theory, as is QCD, which is modeled on QED. Complications come in when dealing with matter fields at the quantum level, but all these can be satisfactorily accounted for by a theory that shows gauge symmetry. But it is one of the crucial features of QED that it is only gauge symmetric because the mass of the photon is zero. If the photon had any mass at all, it would be impossible, it turns out, to renormalize the theory, and we would be stuck with the infinities. This becomes a problem when physicists try to use the successful gauge theory of the electromagnetic interaction as a model for the construction of a similar theory of the weak nuclear interaction, the process that is responsible, among other things, for radioactive decay and the emission of beta particles (electrons) from radioactive nuclei. Just as the electric force is carried, or mediated, by the photon, so it seems that the weak force must be mediated by its own boson. But the situation is more complicated, because in order for electric charge to be transferred during weak interactions, the weak boson (the "photon" of the weak field) must carry charge. So there must actually be at least two of these particles, bosons dubbed W^+ and W^-, and since weak interactions do not always involve charge transfer the theorists had to invoke a third mediator, the neutral Z boson, to complete the set of weak photons. The theory demanded the existence of this particle, initially to the embarrassment of physicists, who had no experimental evidence for its existence.

The correct mathematical symmetries involving the weak interaction, the two W particles*, and the neutral Z were first worked out by Sheldon Glashow, of Harvard University, in 1960, and published in 1961. His theory wasn't complete, but it offered a glimpse of the possibility that one theory might eventually incorporate both weak and electromagnetic interactions. The key problem was that the

---

*The W^+ and W^- can also be considered, of course, as one particle and its antiparticle, like the electron (e^-) and the positron (e^+). In case you aren't confused enough, the W also has another name, the intermediate vector boson.

theory required W particles, unlike the photon, not only to carry charge but have mass, which makes it impossible to renormalize the theory and also breaks the analogy with electromagnetism, where the photon is massless. They must have mass, because the weak interaction has only got a short range—if they were massless then the range would be infinite, like the range of the electromagnetic interaction. The problem isn't so much with the mass itself, however, as with the spin of the particles. All massless particles, such as the photon, are only allowed by the quantum rules to carry their spin parallel or antiparallel to their direction of motion. A particle with mass, like the W, can also carry its spin perpendicular to its motion, and this extra spin state causes all the problems. If the Ws were massless, then there would be a sort of symmetry between the photon and the W, and therefore between the weak and electromagnetic interactions, which would make it possible to combine them into one renormalizable theory explaining both forces. It is because this symmetry is "broken" that the problems arise.

How can a mathematical symmetry get broken? The best example comes from magnetism. We can think of a bar of magnetic material as containing an enormous number of tiny internal magnets, corresponding to individual atoms. When the magnetic material is hot, these tiny internal magnets spin around and jostle one another at random, pointing in all directions, and there is no overall magnetic field to the bar—no magnetic asymmetry. But when the bar is cooled below a certain temperature, called the Curie temperature, it suddenly takes up a magnetized state, with all the internal magnets lined up with one another. At high temperature, the lowest available energy state corresponds to zero magnetization; at low temperature, the lowest energy state is with the internal magnets lined up (it doesn't matter which way they line up). The symmetry is broken, and the change has occurred because at high temperatures the thermal energy of the atoms overcomes the magnetic forces, while at low temperatures the magnetic forces overcome the thermal agitation of the atoms.

In the late 1960s, Abdus Salam, working at Imperial

College in London, and Steven Weinberg, at Harvard, independently came up with a model of the weak interaction, developed from the mathematical symmetry devised by Glashow in the early 1960s and independently by Salam a few years later. In the new theory, symmetry breaking requires a new field, the Higgs field, and associated particles, also called Higgs. The electromagnetic and weak interactions are combined in one symmetrical, gauge field, the electroweak interaction, with massless mediating bosons. This was later shown to be a renormalizable theory, by the work of the Dutch physicist Gerard t'Hooft in 1971, at which point people began to take the theory seriously. With evidence for the Z particle turning up in 1973, the electroweak theory became firmly established. The combined interaction only "works" under conditions of very high energy density, like those in the Big Bang, and at lower energies it is spontaneously broken in such a way that the massive W and Z particles appear and the electromagnetic and weak interactions go their separate ways.

The importance of this new theory can be gauged from the fact that Glashow, Salam, and Weinberg shared the Nobel Prize in Physics for it in 1979, even though there was then no direct experimental proof that their idea was correct. Early in 1983, however, the CERN team in Geneva announced the results of particle experiments at very high energies (achieved by colliding a beam of high energy protons head-on into a beam of high energy antiprotons),

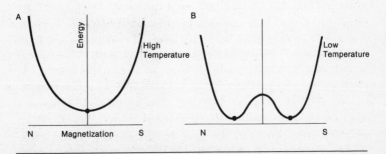

Figure E.5/Symmetry breaking occurs when a bar of magnetic material is cooled.

which are best explained in terms of W and Z particles with masses around 80 GeV and 90 GeV, respectively. These fit the predictions of the theory very well, and the Glashow-Salam-Weinberg theory is a "good" theory because it makes predictions that can be tested, unlike Glashow's earlier theory, which did not. Meanwhile, the theorists have not been idle. If two interactions can be combined in one theory, why not a grand unified theory involving all of the fundamental interactions? Einstein's dream is closer than ever to being realized, in the form not just of symmetry but supersymmetry and supergravity.

SUPERGRAVITY

The problem with the gauge theories, apart from the difficulty of renormalization, is that they are not unique. Just as an individual gauge theory includes infinities that have to be tailored to fit reality by renormalization, so there are an infinite number of possible gauge theories, and the ones chosen to describe the interactions of physics have to be tailored in the same way, on an equally *ad hoc* basis, to fit the observations of the real world. What's worse, there is nothing in the gauge theories that says how many different kinds of particles there ought to be—how many baryons, or leptons (particles in the same family as the electron), or gauge bosons, or whatever. Ideally, physicists would like to come up with a unique theory that requires only certain numbers of certain kinds of particles to explain the physical world. A step toward such a theory came in 1974, with the invention of supersymmetry.

The idea came from work by Julius Wess, at the University of Karlsruhe, and Bruno Zumino, of the University of California, Berkeley. They started from a guess about what things ought to be like in an ideally symmetric world—that every fermion should have a counterpart boson with the same mass. We don't actually see this kind of symmetry in nature, but the explanation could be that symme-

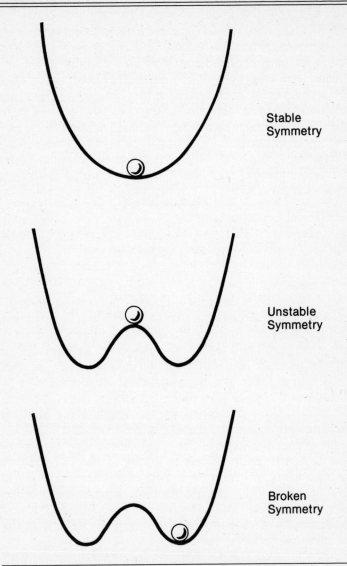

Stable
Symmetry

Unstable
Symmetry

Broken
Symmetry

Figure E.6/The breaking of magnetic symmetry in
Figure E.5 can be understood in terms of a ball in a
valley. With one valley, the ball is in a stable, symmetric
state. If there are two valleys, the symmetrical position
is unstable and the ball must, sooner rather than later,
fall into one valley or the other, breaking the symmetry.

try has been broken, like the symmetry involving the electromagnetic and weak interactions. Sure enough, when you carry the math through, you find that there are ways to describe supersymmetries that exist during the Big Bang but then get broken in such a way that the everyday particles of physics acquire small mass while their super-partners are left with very large masses. The superparticles could then exist only for a short time before breaking up into a shower of less massive particles; to create the super-particles today, we need to create conditions like those of the Big Bang, very high energies indeed, and it will come as no surprise if even the colliding proton/antiproton beams at CERN fail to produce them.

It is all very "iffy." But it has one great plus going for it. There are still different kinds of supersymmetric field theory, variations on the theme, but the restrictions of the symmetry mean that each version of the theory allows the existence of only a definite number of different kinds of particles. Some versions contain hundreds of different fundamental particles, which is a daunting prospect, but others have room for far fewer, and none of the theories predicts the possibility of an infinite number of "fundamental" particles. Even better, the particles are arranged neatly in family groups in each supersymmetric theory. In the simplest version, there is just one boson, with zero spin, and one spin-1/2 partner; a more complicated version has two spin-1 bosons, one spin-1/2 fermion and one fermion with spin 3/2, and so on. But still the best news is yet to come. In supersymmetries, you don't always have to worry about renormalization. In some of these theories, the infinities cancel out automatically, not in an *ad hoc* fashion, following the proper rules of mathematics and leaving sensible finite numbers behind.

Supersymmetry looks good, but it isn't yet the final answer. Something is still missing,. and physicists don't know what it is. Different theories fit different features of the real world quite well, but no single supersymmetric theory explains all of the real world. Nevertheless, there is one particular supersymmetric theory that is worthy of special mention. It is called the $N = 8$ supergravity.

This supergravity starts with a hypothetical particle, called the graviton, which carries the gravitational field. Along with it there are eight more particles (hence the "$N = 8$") called gravitinos, 56 "real" particles such as quarks and electrons, and 98 particles that are involved in mediating interactions (photons, Ws, and many more gluons). This is a formidable number of particles, but it is precisely determined by the theory, with no room for any more. The kind of difficulties physicists face in testing the theory can be seen by considering the gravitinos. These have never been detected, and there are two diametrically opposed reasons why this should be so. Perhaps the gravitinos are elusive, ghostlike particles with very little mass that never interact with anything. Or, perhaps, they are so very massive that our present-day particle machines are inadequate to provide the energy necessary for them to be created and observed.

The problems are immense, but theories like supergravity are at least consistent, finite, and not in need of renormalization. There is a feeling in the wind that physicists are on the right track. But if particle accelerators are inadequate to test the theories, how can they be sure? This is why cosmology—the study of the whole universe—is today a boom area of science. As Heinz Pagels, executive director of the New York Academy of Sciences, said in 1983, "We have already entered the era of post-accelerator physics for which the entire history of the universe becomes the proving ground for fundamental physics."* And the cosmologists are no less eager to embrace the particle physics.

IS THE UNIVERSE A VACUUM FLUCTUATION?

Perhaps cosmology really is a branch of particle physics. For, according to one idea that has progressed over the past

---

*Quoted in *Science*, 29 April 1983, volume 220 page 491.

ten years or so all the way from being thought of as completely crazy to the near-respectability of being regarded
merely as outrageous, the universe and everything in it
may be no more, and no less, than one of those vacuum
fluctuations that allow collections of particles to burst forth
out of nothing, live for a while, and then be reabsorbed into
the vacuum. The idea ties in very closely with the possibility that the universe may be gravitationally closed. A
universe that is born in the fireball of a Big Bang, expands
for a time and then contracts back into a fireball and disappears, *is* a vacuum fluctuation, but on a very grand scale. If
the universe is exactly balanced on the gravitational edge
between indefinite expansion and ultimate recollapse, then
the negative gravitational energy of the universe must precisely cancel out the positive mass energy of all the matter
in it. A closed universe has zero energy overall, and it is not
so difficult to make something with zero energy overall out
of a vacuum fluctuation, even if it is a rather neat trick to
make all the bits expand away from one another and allow,
temporarily, for all the interesting variety we see about us.

I am particularly fond of this idea because I played a
part in its appearance in its modern form in the 1970s. The
original idea can be traced back to Ludwig Boltzmann, the
nineteenth-century physicist who was one of the founders
of modern thermodynamics and statistical mechanics.
Boltzmann speculated that since the universe ought to be
in thermodynamic equilibrium, but manifestly is not, its
present appearance might be the result of a temporary deviation from equilibrium, allowed by the statistical rules
provided equilibrium is maintained, on average, in the long
term. The chances of such a fluctuation occurring on the
scale of the visible universe are tiny, but if the universe
existed in a steady state for infinite time then there would
be a virtual certainty of something of the kind happening
eventually, and since only a deviation from equilibrium
would allow life to exist, it comes as no surprise that we
should be here during a rare departure of the universe from
equilibrium.

Boltzmann's ideas never found favor, but variations on
the theme continued to emerge from time to time. In 1971

the variation that caught my fancy, and which I wrote about in *Nature,* was the possibility of the universe being born in fire, expanding, and then recollapsing into nothing.* Two years later, Edward Tryon, of the City University of New York, submitted a paper to *Nature* developing the idea of the Big Bang as a vacuum fluctuation but referring in his covering letter to my anonymous article as the jumping-off point for his speculations.† So I have a special interest in this particular cosmological model, although it is, of course, quite right that Tryon should now get full credit for bringing in the modern idea of the universe as a vacuum fluctuation. Nobody else thought of it first, but as he pointed out at the time if the universe has zero net energy then the amount of time it is allowed to exist, in line with

$$\Delta E \Delta t = \hbar$$

can be very long indeed. "I do not claim that universes like ours occur frequently, merely that the expected frequency is non-zero," he said. "The logic of the situation dictates, however, that observers always find themselves in universes capable of generating life, and such universes are impressively large."

For ten years the idea was ignored. But recently people have begun to take a new version of it seriously. In spite of Tryon's initial hopes, the calculations suggested that any new "quantum universe" formed as a vacuum fluctuation really would be tiny, a short-lived phenomenon occupying only a small volume of space-time. But then cosmologists discovered a way to make this minuscule universe blossom forth in a dramatic expansion that could make it grow into the size of the universe in which we live in less than the blink of an eye. "Inflation" is the buzzword of cosmology in the middle 1980s, and inflation explains how a minuscule vacuum fluctuation could have grown into the universe in which we live.

---

*Nature, volume 232 page 440, 1971.
†*Nature*, volume 246 page 396, 1973.

INFLATION
AND THE UNIVERSE

Cosmologists were already interested in any extra particles that might exist in the universe, because they are always on the lookout for the "missing mass" needed to make the universe closed. Gravitinos with a mass of about 1,000 eV per particle could be particularly useful—not only would they help to close the universe, but, according to the equations that describe the expansion of the universe out of the Big Bang, the presence of such particles would be just right to form clumps of matter the size of galaxies. Neutrinos with a mass of about 10 eV each would be just right to encourage the growth of clumps of matter on the scale of clusters of galaxies, and so on. But in the past couple of years cosmologists have become even more interested in particle physics, because the latest interpretation of symmetry breaking suggests that the broken symmetry itself may have been the driving force that burst our bubble of space-time out into its expanding state.

The idea came originally from Alan Guth, of the Massachusetts Institute of Technology. It goes back to the picture of a very hot, very dense phase of the universe in which all of the interactions of physics (except gravity; the theory doesn't yet include supersymmetry) were united in one symmetric interaction. As the universe began to cool, the symmetry was broken and the basic forces of nature—electromagnetism and the strong and weak nuclear forces—went their separate ways. Clearly, the two states of the universe, before and after the symmetry breaking, are very different from one another. The change from one state to the other is a kind of phase change, like the change of water into ice when it freezes, or into steam when it boils. Unlike those everyday phase changes, however, the breaking of the symmetry in the early universe should, according to the theory, have generated an overwhelmingly large repulsive gravitational force, blasting everything apart in a fraction of a second.

We are talking about the very early origins of the universe, before about 10^{-35} sec, when the "temperature" would have been above 10^{28} K, in so far as temperature has any meaning for such a state. The expansion produced by the symmetry breaking would have been exponential, doubling the size of each tiny volume of space every 10^{-35} sec. In far less than a second, this headlong expansion would swell up a region the size of a proton to the size of the observable universe today. Then, within that expanding region of space-time, bubbles of what we think of as normal space-time develop and grow through a further phase transition.

Guth's initial version of the inflationary universe did not attempt to explain where the initial tiny bubble came from. But it is very tempting to equate this with a vacuum fluctuation of the kind described by Tryon.

This dramatic view of the universe solves many cosmological puzzles, not least being the very remarkable coincidence that our bubble of space-time seems to be expanding at a rate just on the borderline between being open and closed. The inflationary universe scenario *requires* that just this balance should be struck, because of the relationship between the mass/energy density of the bubble and the inflationary force. More breathtakingly, the scenario consigns us to a very insignificant role in the universe, placing everything that we can see in the universe inside a bubble within a bubble of some much greater expanding whole.

We live in exciting times, apparently poised on the edge of a breakthrough in our understanding of the universe as significant, as Dirac predicted, as the step from the Bohr atom to quantum mechanics. I find it especially intriguing that my search for Schrödinger's cat should have ended up with the Big Bang, cosmology, supergravity and the inflationary universe, because in a previous book, *Spacewarps*, I started out to tell the story of gravity and general relativity and ended up in the same place. In neither case was this originally planned; in both cases supergravity seems a natural end point, and that perhaps is a sign that the unification of quantum theory and gravity may be on the horizon. But there are no neat endings yet, and I hope there never will be. As Richard Feynman has said, "One of the ways of stopping science would be only to do experiments in the region where you know the law." Physics is about probing into the unknown, and:

> what we need is imagination, but imagination in a terrible strait-jacket. We have to find a new view of the world that has to agree with everything that is known, but disagrees in its predictions somewhere, otherwise it is not interesting. And in that disagreement it must agree with nature. If you can find any other view of the world which agrees over the entire range where things have already been observed, but disagrees somewhere else, you have made a great discovery. It is very nearly impossible, but not quite. . . .*

If the business of physics is ever finished, the world will be a much less interesting place in which to live, which is why I am happy to leave you with loose ends, tantalizing hints, and the prospect of more stories yet to be told, each one as intriguing as the story of Schrödinger's cat.

---

*The Character of Physical Law, page 171.

BIBLIOGRAPHY

These are the books that I read during the course of my search for the truth about Schrödinger's cat. I did not set out to provide a comprehensive bibliography of quantum theory, and experts in the field will undoubtedly notice the absence of some titles that they might expect to find here. However, one reference leads to another, and you can find anything of importance that has ever been written about quantum theory, and a lot more besides, by starting out from somewhere in the following selection and following your nose. In addition to the factual texts, I include at the end a selection of science-fiction titles that are not only entertaining but also informative on some quantum themes, especially the idea of parallel worlds.

QUANTUM
THEORY

A. d'Abro, *The Rise of the New Physics*, volume two, Dover, New York, 1951 (original edition 1939).

A comprehensive early treatment for the nonspecialist. Volume one covers the historical and mathematical backgrounds, so volume two is all about quantum theory. The old-fashioned style doesn't make it an easy read for a modern audience, but it is a very thorough treatment (the two volumes together make up 982 pages) well worth a look if you are dedicated enough to work at understanding some of the math.

Kenneth Atkins, *Physics—Once Over—Lightly*, Wiley, New York, 1972.

This is intended as a textbook for a one-semester course in physics for nonscience majors, but it is interesting and clear enough to be of value to the more casual reader. As a serious guide to physics for nonscientists, it is the best of its kind, and carries the reader through from simple beginnings to relativity, quantum mechanics, nuclei, and particles. Although philosophical implications and the meaning of quantum reality are only touched upon, the book provides the basics of quantum cookery clearly for anyone who wants to try putting a few numbers into the equations. Strongly recommended.

Ted Bastin (editor), *Quantum Theory and Beyond*, Cambridge University Press, New York, 1971.
Based on papers from an informal colloquium held in Cambridge in 1968 to consider the possibility that a major "paradigm shift" in quantum theory may be imminent. Mostly heavy going and more philosophical than the majority of the books mentioned here.

Max Born, *The Restless Universe*, Dover, New York, 1951.
The best contemporary account of the new physics by one of the leading figures in the development of quantum theory. Not a history of quantum mechanics, but a "popular" book about physics, including of particular interest one of the first descriptions for the layman of the statistical interpretation for which Born later received the Nobel Prize. Also remarkable for the inclusion, half a century ago, of flip-page cartoons to illustrate dynamic processes.

Max Born, *The Born-Einstein Letters*, Macmillan, London, 1971.
Correspondence between the two great men, with commentaries by Born. Includes several interesting sidelights on quantum theory and Einstein's reluctance to accept the Copenhagen interpretation.

Louis de Broglie, *Matter and Light*, Norton, New York, 1939 (translation of French edition published in 1937; also available as a Dover paperback).
Chiefly of historical interest; an almost contemporary account of the birth of the new physics from one of the participants.

Louis de Broglie, *The Revolution in Physics*, Greenwood Press, New York, 1969.
A not-very-well-translated English version of another much older French book, but also of historical interest.

Fritjof Capra, *The Tao of Physics*, Bantam, New York, 1980.
The first of the new wave of books linking modern particle physics with eastern philosophy, mysticism, and religion. Capra is a physicist and weaves a compelling tale that includes the basic quantum ideas, but not in an historical context.

Jeremy Cherfas, *Man Made Life*, Blackwell, Oxford, 1982.
A straightforward introduction to the mysteries of genetic engineering, its potential, and its limitations.

Barbara Lovett Cline, *The Questioners*, Crowell, New York, 1965.
The story of quantum mechanics told in biographical terms—chapters on Rutherford, Planck, Einstein, Bohr, Pauli, and

Heisenberg. A good read, strong on anecdotal material, but with the minimum amount of physics.

Francis Crick, *Life Itself,* Simon & Schuster, New York, 1982.
An easy introduction to the nature of living molecules, with speculation that life on earth may have arrived from out in the universe at large.

Paul Davies, *The Accidental Universe,* Cambridge University Press, New York, 1982.
A clear but mathematical account of the many cosmic "coincidences" that have led to us being here, including a brief mention of the relevance of the Everett interpretation of quantum mechanics to the anthropic principle. A nonmathematical "popular" account of the anthropic principle is a central theme of the same author's *Other Worlds* (Dent, London, 1980).

Bryce DeWitt and Neill Graham, editors, *The Many-Worlds Interpretation of Quantum Mechanics,* Princeton University Press, 1973.
A collection of reprints of the key papers which established the basis of the many-worlds theory. The book includes Everett's PhD thesis, the 1957 papers by Everett and Wheeler from *Reviews of Modern Physics,* and DeWitt's and Graham's later attempts to extend and popularize the theory, as well as other contributions. A neat one-volume summary of what the fuss is all about.

Paul Dirac, *The Principles of Quantum Mechanics,* Oxford University Press, New York, 1982.
The definitive text, even today, for the serious student. Revised and updated several times, the book includes a section on quantum electrodynamics, and its introductory sections provide as lucid a discussion of indeterminacy, superposition, and the need for quantum mechanics as you will find anywhere. Even if you are not a serious student, it is worth borrowing the book from a library to read the first chapter; if you *are* a serious student, Dirac's approach *from* the mathematics *to* the Schrödinger and Heisenberg interpretations is more logical and intelligible than the way the subject is usually taught today.

Paul Dirac, *Directions in Physics,* Wiley, New York & London, 1978.
Lectures delivered in Australia and New Zealand in 1975. Invaluable as the view of the last surviving member of the band which developed quantum mechanics in the 1920s, and doubly so as direct transcripts of Dirac's entertaining and clear lectures. Includes discussion of ideas such as variable gravity and magnetic monopoles, which highlight the incompleteness of physics today.

Sir Arthur Eddington, *The Nature of the Physical World*, Folcroft Library Editions, Folcroft, Pennsylvania, 1935.

The text of a series of lectures delivered in Edinburgh in 1927, this book provides a rare insight into the impact quantum theory had on one of the great scientists of the 1920s, written at a time when the subject was still rapidly changing. As well as being a leading scientist, Eddington was one of the first, and best, science popularizers.

Sir Arthur Eddington, *Science and the Unseen World*, Folcroft Library Editions, Folcroft, Pennsylvania, 1979.

More lecture material from the same era.

Sir Arthur Eddington, *New Pathways in Science*, Cambridge University Press, 1935.

A series of lectures delivered at Cornell University in 1934. Shows how things had progressed since *The Nature of the Physical World* appeared.

Sir Arthur Eddington, *The Philosophy of Physical Science*, University of Michigan Press, Ann Arbor, 1958 (original edition Cambridge University Press, 1938).

Yet more lectures dating from the late 1930s and, as the title suggests, with more philosophical leanings.

Leonard Eisenbud, *The Conceptual Foundations of Quantum Mechanics*, Van Nostrand Reinhold, New York, 1971.

Uses the minimum of mathematics and stresses the physical significance of the quantum theory—but "minimum" here still means quite a lot. A good guide to the basics that does not go on to explain atomic structure and so on but provides physical and philosophical insight into the puzzles of the quantum world.

Richard Feynman, *The Character of Physical Law*, MIT Press, Cambridge, 1967.

Text of a series of TV lectures, delivered at the Cornell University in 1964 and broadcast on BBC2 in 1965. All very readable, from a master lecturer, and includes a good chapter on the quantum mechanical view of nature.

Richard Feynman, Robert Leighton, and Matthew Sands, *The Feynman Lectures on Physics, Volume III*, Addison-Wesley, Reading, Massachusetts, 1981.

The most approachable textbook introduction to quantum mechanics for serious students. Very good on the famous two-slit experiment, and includes an interesting discussion of superconductivity.

George Gamow, *The Atom and Its Nucleus*, Prentice-Hall, New Jersey, 1961.
An easy read with a fair bit about quanta and wave theory from a master storyteller who happens to have been involved with the story—Gamow worked for a time with Bohr. A little old-fashioned, but fun, and worth investigating just for the sketches of the main characters.

Maurice Goldsmith, Alan Mackay, and James Woudhuysen, editors, *Einstein: The First Hundred Years*, Pergamon, Elmsford, New York, 1980.
A very patchy book that includes an excellent article on Einstein by C. P. Snow.

John Gribbin and Jeremy Cherfas, *The Monkey Puzzle*, Bodley Head, London, and Pantheon, New York, 1982.
A book about human evolution, but including a comprehensive, nontechnical account of the workings of DNA.

Niels Heathcote, *Nobel Prize Winners in Physics 1901–1950*, Henry Schuman, Inc., 1953 (reprinted 1971, by Books for Libraries Press, Freeport, New York).
With brief biographical sketches and summaries of the work for which each prize was awarded, this volume neatly indicates the dominant role of quantum theory in the physics of the first half of the twentieth century. Only two key names are missing—Max Born, who didn't receive his prize until the 1950s, and Ernest Rutherford, who was given his under the heading "chemistry." Worth dipping into.

Werner Heisenberg, *Physics and Philosophy*, Harper & Row, 1959.
The text of a series of lectures delivered at the University of St.Andrews in 1955–56. Includes a brief history of quantum theory and a discussion of the Copenhagen interpretation from one of the founders of quantum mechanics. Completely nonmathematical.

Werner Heisenberg, *The Physicist's Conception of Nature*, Greenwood Press, Westport, Connecticut, 1970 (Harcourt Brace edition published 1958).
Another semiphilosophical volume, worth mentioning here chiefly to ensure that it is not confused with Jagdish Mehra's book with the same name! (See below.)

Werner Heisenberg, *Physics and Beyond*, Harper & Row, New York, and Allen & Unwin, London, 1971.
Subtitled "memories of a life in science" this is anecdotal autobiography with little science but a great deal of insight into Heisenberg the man.

Banesh Hoffmann, *The Strange Story of the Quantum*, Peter Smith, Magnolia, Massachusetts, 1963 (original edition published 1947).

An interesting view of the still relatively new quantum theory from the perspective of the 1940s. The author sometimes falls into the trap of overpopularization, losing the thread of his argument in an attempt to stick to everyday language, but it is still a good read almost forty years after it was written. Worth seeking out if only for the postscript, written in 1959, which lucidly explains developments in the previous decade, including Feynman diagrams and the loss of causality.

Ernest Ikenberry, *Quantum Mechanics*, Oxford University Press, London, 1962.

A book for mathematicians and physicists, not a layman's guide. Strong on the "how to" use of quantum theory to solve problems, but weak on interpretation of what the equations mean.

Max Jammer, *The Conceptual Development of Quantum Mechanics*, McGraw-Hill, New York, 1966.

A very comprehensive one-volume study, which pulls no mathematical punches but from which you can still get a lot of interesting insights even if you skip most of the math.

Max Jammer, *The Philosophy of Quantum Mechanics*, Wiley, New York & London, 1974.

A book about the interpretation of quantum mechanics and its philosophical significance. Sometimes stifling detail about the history of, say, the Copenhagen interpretation, but goes far beyond the recipes of quantum cookery.

Pascual Jordan, *Physics of the 20th Century*, Philosophical Library, New York, 1944.

Like the books by de Broglie mentioned above, this is chiefly of historical interest as an account by one of the leading inventors of twentieth-century physics.

Horace Judson, *The Eighth Day of Creation*, Simon & Schuster, 1982.

This large, somewhat ramshackle book about the revolutionary development of molecular biology in the second half of the twentieth century is well worth reading in its own right, both for the story of molecular biology and for the insights into how scientists work. Its particular relevance to the story of the quantum revolution is the clear way in which Judson emphasizes that the birth of what we now call molecular biology occurred when Linus Pauling applied the rules of quantum mechanics to develop an understanding of the chemistry of complex molecules. Unfortunately, Judson also says, incorrectly, that the Heisenberg, Born, and Dirac versions of quantum mechanics appeared after Schrödinger's—but nobody is perfect.

Jagdish Mehra (editor), *The Physicist's Conception of Nature*, Kluwer, Boston, 1973.
The proceedings of a symposium held in Trieste in 1972 to honor the seventieth birthday of Paul Dirac. An amazing list of contributors—reading like a *Who's Who* of quantum theory—makes this epic 839-page volume one of the best guides, for the scientifically literate, to the way in which physics was transformed in the twentieth century.

Jagdish Mehra and Helmut Rechenberg, *The Historical Development of Quantum Theory*, Springer-Verlag, New York, 1982.
This is the definitive historical study of quantum physics. Four volumes published so far take the story up to 1926, and a further five volumes are planned to bring it up to date. Although the epic work pulls no mathematical punches, the many equations in it are surrounded by a wealth of highly readable information.

Abraham Pais, *Subtle Is the Lord . . .*, Oxford University Press, London & New York, 1982.
The definitive account of Einstein's life and work.

Heinz Pagels, *The Cosmic Code*, Simon & Schuster, New York, 1982.
A valiant attempt to explain relativity theory, quantum theory, and modern particle physics in one volume. The heart of this book, by a particle physicist, is the detailed account of the particle zoo—quarks, gluons, and all the rest. Quantum theory is presented here more briefly, as the necessary background to understanding the particles in the zoo, and there is no historical perspective. A good place to go if you want to know more about the proliferation of particles. The book also makes an interesting comparison with those of Capra and Zukav.

Jay M. Pasachoff and Marc L. Kutner, *Invitation to Physics*, W. W. Norton, New York & London, 1981.
Although ostensibly a textbook for nonscience majors, this book provides an accessible overview of the whole of physics with very little mathematics and can safely be recommended to anyone with an interest in modern science.

Max Planck, *The Philosophy of Physics*, W. W. Norton, New York, 1963 (original edition 1936).
Of historical interest only—but an insight into the thinking of the man who, without originally appreciating the enormity of the step he had taken, founded the quantum theory of radiation.

Erwin Schrödinger, *Collected Papers on Wave Mechanics*, Chelsea
    Publishing Company, New York, 1978 (translation of German
    edition published in 1928).
    The basic papers in which Schrödinger laid the foundations of
    wave mechanics, including his analysis that demonstrated the
    equivalence of matrix and wave mechanics. The essential orig-
    inal papers on matrix mechanics have been collected by van der
    Waerden (see below).

Erwin Schrödinger, *What Is Life?*, Cambridge University Press, New
    York, 1967 (original edition 1944; this edition combined in one
    volume with *Mind and Matter*, originally published 1958).
    A beautifully written book of historical interest as a major influ-
    ence on the people who unraveled the structure of living mole-
    cules. Still worth reading, although it is now known that the
    molecule of life is DNA, and that genes are not made of protein,
    as Schrödinger assumed when he wrote this book. If this doesn't
    convince you quantum theory is of key importance to genetic
    engineering, nothing will.

Erwin Schrödinger, *Science, Theory and Man*, Dover Publications/
    Allen and Unwin, London, 1957 (original edition 1935).
    Includes Schrödinger's Nobel address, which is clear, infor-
    mative, and essential reading for anyone interested in the de-
    velopment of quantum mechanics.

Erwin Schrödinger, *Letters on Wave Mechanics*, Philosophical Li-
    brary, New York, 1967.
    Letters to and from Schrödinger, the other correspondents being
    Einstein, Planck, and Lorentz. Intriguing historical insights into
    the minds of these great men, including some key correspon-
    dence on the famous cat paradox.

John Slater, *Modern Physics*, McGraw-Hill, New York, 1955.
    A book written with the minimum of mathematics, but for se-
    rious students. In spite of its age this is an excellent introduction
    to quantum theory at the undergraduate level.

J. Gordon Stipe, *The Development of Physical Theories*, McGraw-Hill,
    New York, 1967.
    A basic introduction at freshman university level which, unlike
    so many books for that audience, includes a good introduction to
    quantum theory and nuclear physics. A textbook, not a layman's
    guide.

B. L. van der Waerden (editor), *Sources of Quantum Mechanics*, Peter
    Smith, Magnolia, Massachusetts, 1967
    A collection of the essential original papers, all in English-
    language versions, leading up to, and including, the papers that

laid the foundations of matrix mechanics (Heisenberg, Born, Jordan, and Dirac)—but *not* including Schrödinger's wave mechanics (collected separately; see Schrödinger). Concise but comprehensive introductions to each paper put the work in perspective.

James D. Watson, *The Double Helix,* Atheneum, New York, 1968.
A racy, vivid personal account of the discovery of the structure of DNA. Not so much "warts and all" as "*only* warts"; but great fun and well worth reading.

Harry Woolf (editor), *Some Strangeness in the Proportion,* Addison-Wesley, Reading, Massachusetts, 1980.
This book presents the proceedings of a symposium held at the Institute for Advanced Study, Princeton, to mark the centennial of the birth of Albert Einstein. The list of contributors reads like a *Who's Who* of theoretical physics, and there is a comprehensive section on Einstein's contribution to quantum theory. Although largely nonmathematical, some of this is quite deep and not for the casual reader.

Gary Zukav, *The Dancing Wu Li Masters,* Bantam, New York, 1980.
In effect this book is a counterpart to Capra's *The Tao of Physics,* telling the same story from the viewpoint of someone who is *not* a trained physicist. All scientists should read this, to find out what nonscientists make of the new physics; nonscientists are cautioned that Zukav sometimes lets his excitement get the better of him, that the science in the book is not always 100 percent accurate in its portrayal, and that, like Capra, he pays scant attention to the way the ideas developed. But it is still a good read.

SCIENCE FICTION

Gregory Benford, *Timescape,* Pocket Books, New York, 1981.
The best portrayal in science fiction of what it is like to be a research physicist, combined with a superb fictional portrayal of the kind of time travel that may be possible in a many-worlds reality.

Philip Dick, *The Man in the High Castle,* Gregg Press, Boston, 1979.
A parallel-reality story set in a world where the United States lost World War Two. Nicely written with minimal science but a slight twist that takes it out of the rut.

Randall Garrett, *Too Many Magicians*, Ace Books, New York, 1981.
"What if" stories set in a parallel reality where Richard Lionheart survived for long enough to ensure that the succession to the English throne did not pass through his brother John. Scientifically slight, but good detective stories, and fun.

David Gerrold, *The Man Who Folded Himself*, Amereon, Ltd., Mattituck, New York, 1973.
A funny and entertaining portrayal of the confusing effects of travel forward and backward in time among the many worlds of perpendicular reality. It's easy to dismiss the "science" in this as hocus-pocus, but the implications are very close to some of the ideas spelled out in Chapter Eleven of the present book.

Keith Roberts, *Pavane,* Hart-Davies, London, 1968 (paperback Panther).
Perhaps this story is set in a parallel universe, perhaps not. Either way it makes good reading.

Jack Williamson, *The Legion of Time*, Sphere, London, 1977.
First published as a magazine serial in 1938, this is a competent action adventure story in the SF tradition of its day, remarkable for only one thing. As far as I have been able to trace, this was the first time, in fact or fiction, that the concept of parallel worlds, later to become the many-worlds interpretation of quantum mechanics, appeared in print. Of course, there are older "what if" stories set in alternative realities, but Williamson used respectable scientific language to set his scene, only a decade after the foundations of quantum mechanics had been laid. "Geodesics have an infinite proliferation of possible branches, at the whim of subatomic indeterminism." Hugh Everett, in his doctoral thesis nineteen years later, couldn't put it any more succinctly, although he did put it on a secure mathematical footing. It's seldom that SF really does anticipate the advance of theoretical science, and well worth noting when it does happen.

Robert Anton Wilson, the *Schrödinger's Cat* trilogy *(The Universe Next Door, The Trick Top Hat, The Homing Pigeons)*, all published by Pocket Books, New York, 1982.
It is almost impossible to describe this funny, irreverent, and brilliant trilogy in which three different variations on the quantum theme (one in each volume) are applied with scrupulous care to provide the framework for more or less the same set of actions involving more or less the same set of characters. In a way, the *Schrödinger's Cat* trilogy does for quantum theory what Lawrence Durrell's *Alexandria Quartet* did for relativity theory—but Wilson is funnier. An acquired taste, but if you can acquire it you will have the true flavor of the quantum world on your tongue.

SF writers are constantly "discovering" quantum theory, and every few months a new short story appears from someone who has just latched on to the possibilities. Recent examples are Greg Bear's "Schrödinger's Plague," from *Analog*, 29 March 1982, and Rudy Rucker's "Schrödinger's Cat," from *Analog*, 30 March 1981. There are other stories as good, but I mention these two because it was their use of Schrödinger's cat as a device to grab the attention of an audience unfamiliar with quantum theory that set me off on the trail of revision and discovery that led to the present book, and which gave me my title. My thanks to both authors and to Stan Schmidt, *Analog*'s editor.

INDEX

A

Accidental Universe, The (Davies), 252*n*, 281

"Actions," 43

"Actuarial tables," 62, 65–66

Albert Einstein: Philosopher Scientist (ed. Schulpp), 22*n*

Alexandria Quartet (Durrell), 288

Alpha radiation, 61, 70, 130, 131; Rutherford and, 29–31

Analog (magazine), 289

Anderson, Carl, 126

Angular momentum, *see* Spin

Annalen der Physik (journal), 22, 23

Annales de Physique (journal), 87

Anthropic principle, 252, 281

Antineutron, 127

Antiproton, 127

Aristotle, "four elements" of, 19–20

Arrow of time, 159–60, 190

Aspect, Alain, 3, 227, 228

Aspect experiment, 3–4, 77–78, 213, 227–29, 231, 232, 253

Atkins, Kenneth, 279

Atom and Its Nucleus, The (Gamow), 282–83

Atomic bomb, 132

Atomic decay, *see* Radioactive decay

Atomic energy, *see* Nuclear technology

Atomic mass unit, definition of, 69–70

Atomic number (Z), definition of, 69

Atoms, 1, 19–32; ancient Greeks and, 20, 26; chemistry and, 70–77; de Broglie's image of, 88; discovery of electrons, 24–25; discovery of ions, 25–27; discovery of radioactivity and, 27–29; Einstein's introduction of probability, 61, 62, 64–66, 83, 100, 134; Einstein's proof of existence of, 22–23, 48; emergence of modern concept of, 20; in hyperfluidity, 146; introduction of Bohr's model of, 52–53, 56–66, 93; laser research's debt to Bohr's model of, 134, 137; light's interaction with (*see* Blackbody radiation; Photons; Photoelectric effect); in mature quantum theory, 88, 92, 123 (*and see* Wave/particle duality; Wave mechanics); modern perspective on Bohr's model of, 66–70, 92; nineteenth-century views of, 20–22; orbital analogy and, 33–34, 52–53, 92; radiation from, and cold future of universe, 189–90; Rutherford's nuclear model of, 31–33, 51, 68–69; size of, 67–68, 70; Thomson's model of,

ABOUT THE AUTHOR

After completing his PhD in astrophysics at the University of Cambridge, JOHN GRIBBIN worked for five years on the editorial staff of the journal *Nature,* chiefly responsible for the daily Science Report in the *Times.* He left in 1975 to join the Science Policy Research Unit of the University of Sussex, working in the "futures" team on a study of the likely impact of climatic change on world food supplies. A book about the work of SPRU—*Future Worlds*—was the main result of this three-year stint. Since 1978, John Gribbin has been Physics Consultant to the weekly magazine *New Scientist.*

His 27 books published to date range from astronomy through geophysics and climatic change to human evolution, and include two novels written jointly with Douglas Orgill. They included *White Holes* (1977), *Timewarps* (1979), *Genesis: The Origins of Man and the Universe* (1981) and *The Monkey Puzzle* (co-author with Jeremy Cherefas, 1982): the novels are *The Sixth Winter* (1979) and *Brother Esau* (1982). As well as being a consultant to *New Scientist,* John Gribbin contributes to the "Futures" section of the *Guardian* and to broadcasts on the BBC World Service, British Forces Radio, and occasionally on the BBC's domestic services. His interest in the puzzle of gravity and warped spacetime goes back to his student days in Cambridge, when he won two awards for work on these themes.

Married with two sons, John Gribbin was born in 1946 and lives in Lewes.